Energy
Today and Tomorrow

ENERGY
TODAY AND TOMORROW
Living
with Uncertainty

Joel Darmstadter
Hans H. Landsberg
Herbert C. Morton
with **Michael J. Coda**

A BOOK FROM RESOURCES FOR THE FUTURE

Prentice-Hall, Inc.
Englewood Cliffs, New Jersey 07632

Library of Congress Cataloging in Publication Data

DARMSTADTER, JOEL (date)
 Energy, today and tomorrow.

 "A book from Resources for the Future."
 Includes bibliographies and index.
 1. Power resources. 2. Energy conservation.
 3. Energy consumption. I. Landsberg, Hans H.
 II. Morton, Herbert Charles. III. Resources for the
 Future. IV. Title.
 TJ163.2.D37 1983   333.79   83-8644
 ISBN 0-13-277640-5
 ISBN 0-13-277632-4 (pbk.)

Editorial/production supervision
 and interior design: Kathryn Gollin Marshak
Cover design: Photo Plus Art—Celine Brandes
Manufacturing buyer: Anthony Caruso

Printed in the United States of America

10 9 8 7 6 5 4 3 2 1

ISBN   0-13-277632-4 {P}
ISBN   0-13-277640-5 {C}

Prentice-Hall International, Inc., *London*
Prentice-Hall of Australia Pty. Limited, *Sydney*
Editora Prentice-Hall do Brasil, Ltda., *Rio de Janeiro*
Prentice-Hall Canada Inc., *Toronto*
Prentice-Hall of India Private Limited, *New Delhi*
Prentice-Hall of Japan, Inc., *Tokyo*
Prentice-Hall of Southeast Asia Pte. Ltd., *Singapore*
Whitehall Books Limited, *Wellington, New Zealand*

Contents

CAT Jun 28 '84

5-1-84   DMLS   13-27

83-7980

List of Tables,
Figures, and Plates

TABLES

FIGURES

PLATES

Foreword

This book was written in the belief that there is a clear need for a broad and balanced introduction to the assessment of energy issues. A decade has passed since the major petroleum-exporting countries radically changed the global system of oil production and pricing. The economic costs and political consequences have been far-reaching, and the story is far from over. Having survived one decade's dislocations, we need to think about how to live with the energy problem in the years ahead.

Most books about energy are written from a given vantage point, and this one is no exception. It is written from an economic perspective. Although the book takes into account technological, political, and social factors, it places particular emphasis on economic ones—that is, how prices govern the availability, use, and conservation of energy resources. The authors are especially concerned with the role of the marketplace in facilitating energy choices and in determining how these choices affect economic growth, environmental integrity, national security, and other social goals. But limitations of the marketplace also lead them to consider the role of government.

A book by physicists, biologists, engineers, or political scientists would have a different emphasis. But from whatever perspective a book about energy might be written for the general reader, it would cover certain fundamental issues. Darmstadter, Landsberg, and Morton have tried in this volume to identify and analyze these key issues and to be fair and balanced in their treatment of them.

To write from an economic perspective does not imply that all economists see the issues treated here in the same light. Even among the authors there are bound to be disagreements—as there are differences among scientists about the safety of nuclear energy or the likelihood that an increase in the use of coal will eventually alter the global climate.

On a subject as vast as energy—as controversial and as plagued by gaps in knowledge—it would be foolish to assert that there can be only one authoritative discussion. The authors have tried to make their approach to the topic a good one, but they want their readers to understand that there is more ground to cover. We hope that readers will be encouraged to do their own thinking on the debatable issues treated here and to push beyond the limits of this volume to explore the problem of energy in greater depth.

The structure of the book is simple. Chapter 1 describes why energy has become one of society's major concerns. It highlights a limited number of important "givens" about energy that need to be kept in mind if one is not to be misled by today's headlines. It next looks at U.S. energy issues from the vantage point of different segments of the population; and it winds up with a brief discussion of the U.S. energy situation and outlook at the beginning of the 1980s.

Chapter 2 discusses the many ways in which we use energy, both as individuals and collectively, and highlights what it means to use energy efficiently—that is, in a conserving way. The chapter reviews energy-efficiency achievements and future prospects. It also looks at the way in which rising energy prices—which spur increases in efficiency—affect the nation's overall economic performance. Chapter 3 looks at resources, and what we mean, or should mean, by the term *adequacy*. The energy resources of the United States are described in some detail and placed in a global context, both for conventional and novel sources or forms of energy. But with continuing large inroads, depletable resources do indeed deplete. How science and technology, or research and development, come to the rescue by expanding our capacity for doing old things better and supplementing them by new ways is the burden of chapter 4.

Technology alone, of course, is of little use; nor do we benefit much from resources in the ground, unless institutions provide an efficient system for moving them from producers to consumers. The market mechanism and the government actions that support and supplement it are the topic of chapter 5. A major focus of that chapter is the public policy debates which have centered on the effectiveness of competition in energy industries and the pros and cons of government regulation.

But no matter what mix of private–public sector arrangements prevail, there is, as the cliché reminds us, "no free lunch." That is, energy activities unavoidably leave an unwanted impact on our physical environment. Chapter 6 looks at the environmental constraints that put limits on our actions; these constraints affect all phases of the energy cycle—from searching for resources to utilizing them and disposing of the residual waste.

In chapter 7 the authors extend their horizon beyond our national borders. Whatever we do at home affects the rest of the world, and vice versa. The price we pay for oil is shaped by events and decisions abroad. The fall or rise of a political regime in a distant country can wreak havoc with our energy economy; nuclear weapons proliferation is a constant threat that can be lessened by internationally agreed safeguards governing civilian nuclear power programs; supply problems of our allies cannot but have their effect on this country's policies. The scope and sense of the chapter is conveyed by its title—"Energy in an Unstable World." Finally, in an Epilogue, the writers attempt to draw some highly simplified profiles of major energy issues, describe divergent perceptions and attitudes, and formulate the kinds of criteria and questions that nonexperts might want to keep in mind as they attempt to follow the continuing debate in the years ahead.

For thirty years, Resources for the Future has played an active role in research and education on natural resources and environmental issues. Its attention to clear-headed analysis and accuracy and its unwillingness to take an advocacy stance have made RFF and its publications an important source of knowledge and influence in both the classroom and policy arena.

Although numerous RFF publications continue to emphasize comprehensive analytical treatment of resource topics, we have also moved increasingly to adapt the results of our scholarship and research in a way as to make them accessible to a wide, rather than just specialized, audience. The present book sustains RFF tradition in its insistence on objectivity and accuracy; it builds on that tradition by a conscious effort—in style, organization, and content—to expose a broad segment of the reading public to one of the major issues of the day.

Washington, D.C.
May 1983

Emery N. Castle, President,
Resources for the Future
Milton Russell, Director
of RFF's Center for Energy
Policy Research

Acknowledgments

We invited criticism on a draft of this book from persons in industry, research and education, and the press. The resulting comments from the following individuals proved extremely helpful in the preparation of this final version: John Day, Theodore R. Eck, John M. Fowler, Robert Fri, and Colin Norman. We express thanks as well to our RFF colleagues—Harry Broadman, Glen Gordon, Harry Perry, Paul Portney, and Milton Russell—for their helpful remarks on the earlier version.

We also field-tested that draft as a text for students of Ian Barbour's Science, Technology, and Public Policy Program at Carleton College in Northfield, Minnesota. It is hard to imagine a group of guinea pigs responding as thoughtfully and articulately as did this group in their evaluation of the manuscript. Both the students and their teacher contributed to what we hope is an improved book for their effort.

We bear sole responsibility, of course, for the final product. Routine as that disclaimer may be, it is worth emphasizing here. Convictions on energy are sometimes quite divergent; and the final text can reflect many, but never all, of the judgments expressed by reviewers.

Elizabeth Davis did a conscientious job of verifying data and offered constructive editorial suggestions. Lee Carlson, Elisabeth Hale, and Margaret

Parr-Recard skillfully handled the typing of successive revisions. And Jo Hinkel's proven capabilities as an RFF editor once again resulted in touches of grace and coherence that otherwise would have eluded us.

J.D.

H.H.L.

H.C.M.

M.C.

Energy
Today and Tomorrow

CHAPTER 1

Introduction
and Major Themes

If one were to search the card catalog of a major library under the entry "Energy," one would discover that by far the largest number of books, pamphlets, and articles on the subject have been published since 1974. There is an easy explanation. When, in October 1973, the Arab oil-exporting countries embargoed the shipment of oil to the United States and several other countries, and when three months later the countries which had earlier banded together in the Organization of Petroleum Exporting Countries (OPEC) raised the price of oil from a little over $3.00 a barrel to nearly $12.00, these actions were more than a sudden storm that would blow over. It was a profound sea change. In earlier years, energy availability, costs, policy, demand, and trade had been topics of interest only to a small group of insiders and had made headlines only rarely. Now they had become matters of great moment to every consumer as well as to the expert. They could no longer be taken for granted and remain in the backwaters of public concern.

There had been anxious moments before. The Suez Canal, a major passageway for Persian Gulf oil, was closed following the war of 1956. In 1967 Saudi Arabia and several other Arab oil exporters embargoed oil shipments to the United States. But these were only "moments." U. S. oil production was still expanding, the United States had spare capacity for producing oil that could replace shipments from the Middle East, and oil demand had not yet experienced the massive worldwide growth that occurred in the decade preceding 1973. After some anxiety, life had returned to normal.

This was not so after 1973. By then, responding to rapidly rising demand, world oil output had more than quintupled, from 3.8 billion barrels in 1950 to 20.4 billion barrels, rising by 1 billion barrels or more a year after 1965 (Figure 1-1). Not only had U. S. oil production peaked in 1970, but there was no spare oil capacity in the United States or elsewhere outside the OPEC-affiliated countries.

Moreover, the 1960s saw a new entry appear on the public agenda: "Environment." Both at home and abroad concern had arisen that the capacity of land, water, and air to absorb pollutants and withstand onslaughts on the quality of the human environment was being exhausted. Most sources of energy are major polluters, as is set out in Chapter 6; thus proposals to increase domestic production of fossil fuels and nuclear power met with growing opposition. Making the situation even more difficult, people and machines were geared to energy abundance at low cost. Even if users were able, they were not sufficiently motivated to move quickly toward using energy more efficiently, or, as it is usually put, toward conserving energy.

The shock of being at the mercy of an association of foreign suppliers, most of them thousands of miles away, was both severe and slow to be recognized as heralding long-term change. What was happening was generally referred to as the "energy crisis," as if it were an event that would soon be

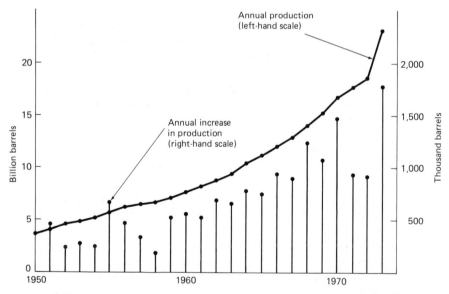

Figure 1-1   The growth of world crude oil production in billions of barrels, from 1950 to 1973. Production increased steeply beginning in the early 1960s. Toward the end of the period annual increments were more than double those at the beginning. [Data from U.S. Department of the Interior, *Energy Perspectives* (Washington, D.C., June 1976).]

resolved and followed by a return to normalcy. Normalcy, of course, was cheap, abundant, reliable energy.

The events of 1973-74 had a severe impact on the economy. The sudden, steep rise in oil prices, following the troublesome but short-lived embargo, depressed the rate of economic growth and accelerated inflation. It created severe balance-of-payments problems, as the oil exporters exacted vast amounts of money from the oil importers, but were without well-laid plans to quickly spend these funds. Thus, as if hit by a sudden, large excise tax, domestic demand in the oil-importing countries was depressed, while prices rose. This development further aggravated the "stagflation" problem—that is, the simultaneous existence of inflation and slow growth—that had already begun to plague the economies of the industrialized nations. Energy, once the engine of economic growth, had become an impediment, a powerful brake.

Nor was this all. Rising energy prices affected different people and groups with different severity. Those who depended heavily on cheap, reliable energy were hit hardest. They included people living in suburbs and others dependent on private automobiles, people in cold climates, those making their living in the recreation and tourist trades, and all those involved in the production of automobiles, appliances, and other energy-consuming machines. Those living in the northeastern United States who relied heavily on fuel oil saw their home-heating costs rise steeply; energy consumers elsewhere, who were served mainly by price-regulated natural gas, enjoyed a measure of protection, at least for most of the 1970s. Similarly, certain consumers of electricity, especially in some of the western states and in other areas served heavily by cheap hydroelectricity, escaped the full brunt of the price increase. As a result, a mood of divisiveness emerged, pitting region against region and occupation against occupation. It hardly had the makings of another civil war, as some contemporary rhetoric would have it, but it was an important element in thwarting the public consensus so important for policymaking.

Quickly, the steep increase in the price of oil directed attention to competing sources of energy. While speculation about "running out" of oil was not new—as early as 1920 the director of the U. S. Geological Survey had written that U. S. oil production must peak within five and possibly only three years—it now seemed more convincing. At the level of consumption reached at the beginning of the 1970s vast new discoveries would be required to meet future demand, and it was becoming increasingly difficult to imagine where these new supplies might be found. Without such new flows, it was generally assumed that oil producers would have little trouble in unilaterally setting the supply and price of the resources they controlled and that the major escape hatch for oil-importing countries was a turn to alternative sources of energy. Some of these would substitute directly for oil and gas and others would do so in more roundabout ways, often involving costly changes in equipment.

These alternative resources were coal and nuclear in the first instance—even though neither could directly replace oil in transportation and both had substantial, some would say fatal, flaws; oil and gas substitutes made from coal, shale, and tar sands in the intermediate future; nondepletable sources, such as solar energy, biomass (crops and trees), the ocean, the wind, and so on in the longer run; and possibly nuclear fusion on a distant horizon. Thus, the decade of the 1970s saw a great expansion of energy research and development designed to bring into being new energy sources and processes and lead to a greatly different energy mix in the future.

A final sign of the changing times was the concern over having enough fuel for emergencies—the search for what was called "security of supply." A second oil shock, on a smaller scale than that of 1973–74, took place in 1979–80, in the wake of the Iranian revolution. Exports of Iranian oil declined sharply, consuming countries built up their oil stocks to guard against future scarcity, and OPEC once again quickly and steeply boosted the price of its oil. The event confirmed the need for measures that would provide a buffer against future supply cutbacks, a breathing space in the event of another severe price boost, and perhaps even a deterrent to potential market manipulators. The U. S. strategic petroleum reserve, that had been languishing for years after it was first legislated by the Congress in 1975, was the major vehicle. By late 1982, it had reached a near-respectable level of close to 300 million barrels—equal to about sixty days of imports and worth about $10 billion on the market. Together with other security approaches, described in Chapter 7, it heralded the arrival of a wholly new energy environment in the United States.

CHALLENGES AND REALITIES

It is tempting to try to capture the essence of the new energy situation in a few sentences or propositions, but we know that the energy situation is subject to rapid change. Thus, in 1981, while the United States was still struggling to adjust to a shortage and high prices, an "oil glut" emerged, culminating in March 1983 in the first official OPEC price cut. All this in the face of a still greatly diminished oil output in Iran and Iraq! Anyone suggesting in the early seventies that by 1981 oil would go begging in spite of Iranian and Iraqi supplies being reduced to a fraction of their former volume would have been declared a maverick or eccentric. As it turned out, conventional wisdom was in error. Still, the temptation remains to search for a few plausible propositions that are not likely to be invalidated in a matter of two or three years. One way of putting it is that the world faces the following challenges:

- To adjust to high and, in the longer run, rising real energy prices
- To reduce its dependence on oil and gas and diversify its sources of energy supply, turning first toward coal, and perhaps nuclear power, and—

The Nature of OPEC

Formed in 1960, by five oil-exporting countries and devoted primarily to keeping the international oil companies from lowering the posted prices that were the basis for calculating the exporters' revenues, the Organization of Petroleum Exporting Countries (OPEC) led a quiet life until, having grown to thirteen members, it jumped into the headlines in late 1973 and became known as the organization that quadrupled the price of oil in less than a year. Since then OPEC has been called all sorts of things, but most frequently a "cartel." Now a true cartel is a compact of suppliers of a given commodity that endeavors to set the level of supply, apportion sales territories, and establish the price. In simple terms, a cartel "rigs the market." It can do so successfully when its members exercise control over the bulk of production, when there is little chance that output will expand outside the cartel, and when consumers will be for all practical purposes locked into buying the particular product at accustomed rates.

When OPEC emerged as an operating force in 1973, conditions for success seemed to be right. To be sure, consumers were not that tightly locked in, control was not that all-embracing, and there was the potential for production growth outside OPEC. Moreover, OPEC never attempted to assign exclusive sales territories. It never was a textbook cartel. Nonetheless, its members collectively managed (1) to control output so that no significant slack would develop, and (2) to maintain the price set for that output. They could not, however, prevent gradual additions to oil production by non-OPEC exporters, most prominently Mexico and the United Kingdom.

Does it matter what we call OPEC: a cartel, a quasi cartel, a price leader? Not really, provided we understand the context in which it operates, its sources of strength, and its limitations. Its strength is its ability to sustain a given level of prices by adjusting output to it. This requires some kind of organized, collective action. Its limitations lie in the gradual erosion of market control as consumers "unlock" themselves by conserving oil and replacing it with other fuels in order to reduce demand, and as the more aggressive members get restless in times of slack demand, and as new suppliers emerge that operate outside OPEC all the while benefiting from the protection afforded by the price umbrella held by OPEC over the entire market.

In summary, it is not so simple as deciding to set the price. To do so successfully, conditions of demand and supply have to be right and stay right. Time will tell whether OPEC members' agreements to set export quotas and lower the price, hammered out in March 1983, is only a passing phase in the unmaking of OPEC or proof of its durability.

at the rate favored by affordable technology and spread of information—eventually toward renewable sources

- To accomplish these first two steps with the least damage to the growth
of the economy (including equitable sharing in the growth), the environment, and international order
- To provide contingency plans for surprises and rude shocks that are sure
to occur but whose specific nature and timing cannot be predicted.

A second way of characterizing the situation is to capture it in a few
grossly simplified "realities," as was done in 1979 by an expert study group
in a book entitled *Energy—The Next Twenty Years.** These realities appear
still valid and are here briefly reiterated and extended.

Reality one. "The world is not running out of energy." That is to
say, there is a great abundance of physical energy resources, even though the
handiest and cheapest have been used first. What matters is the cost at which
they become available and the way the world goes about bringing them into
production. As we shall see in Chapter 3, which deals with this reality, even
of oil there are vast deposits that could be exploited, if "the price is right."

Reality two. "Middle East oil holds great risks, but is so valuable that
the world will remain dependent on it for a long time." The risks, of course,
are supply interruption, price boosts, and, on the political side, a possible
power struggle over domination of these resources. Nonetheless, Middle East
oil will be around for quite a while, certainly beyond the end of this century.
Both Chapters 3 and 7 bear on this subject.

Reality three. "Higher energy costs cannot be avoided, but can be
contained by letting prices rise to reflect them." OPEC is only a part of this
piece of wisdom. Even without it, bringing new sources into the supply stream
will be costly. Oil now comes from more remote places, is sought in hostile
environments (the North Sea or the Arctic, for example) and under technologically more complex conditions, and requires heavy capital investment. But
to the extent that we let energy prices reflect full costs—including the cost of
environmental protection—we will also stimulate greater efficiency in use (as
is shown in Chapter 2) and a nurturing of competing sources, as indicated in
the discussion of market structure (see Chapter 5). There is no alternative to
letting the higher costs be accommodated.

Reality four. "Environmental effects of energy use are serious and
hard to manage." This is the theme of Chapter 6. It is a harsh fact of life that

---

*Hans H. Landsberg (ed.), *Energy: The Next Twenty Years* (Cambridge, Mass., Ballinger,
1979).

one cannot produce, transport, or consume energy without creating a pollutant. The two go hand in hand. The best we can do is favor those sources that pollute least and engineer technologies that capture, neutralize, and dispose of the noxious substances safely. That is not always possible, or at least not yet. Meanwhile, the costs of the admittedly less than perfect approaches must enter the price at which energy is sold.

Reality five. "Conservation is an essential source of energy in large quantities." Of course, conservation, which means using energy more efficiently, is not really a "source," as is coal or oil, just as buying a television set at a large discount is not a "source" of income. That is why we put the word in quotes. But in the larger scheme of things, conservation enables us to do the same things we would otherwise have had to do by consuming added units of energy. So we will not quibble over the use of the word. What is more important is that over the past decade by using energy more efficiently and switching from activities that use more to those that use less energy, we have made one of the most significant conservation efforts in the energy field (see Chapter 2). And experts agree that there is a lot of steam left in the conservation engine. One need only realize that the average automobile in use today in the United States gets little more than 15 miles per gallon. The gas-guzzlers are still out there, but their impact will be continuously declining, as they are replaced by the 30- and 40-miles-per-gallon automobiles. Still more efficient machines are no longer a dreamer's product, and for the low-mileage automobiles produced, every company must turn out enough high-mileage ones to meet the legislatively mandated average efficiency standard that will reach 27.5 mpg in 1985.

Reality six. "Serious shocks and surprises are certain to occur." This single statement should really continue with "but nobody can tell what they will be and when they will occur." In 1975 no one predicted the Iranian revolution or the Iranian–Iraqi war, or the Three Mile Island accident. But everybody realized that the Middle East was a highly unstable part of the world and that any political disturbance would affect the supply of oil; and that nuclear power generation was a hazardous technology subject to inevitable breakdowns and the possibility, however remote, of a severely damaging accident. What will happen next? Nobody knows, except most expect the surprises to be unpleasant ones. The Middle East is hardly more stable than before; the public continues to feel uncomfortable with nuclear power, as aberrations of various kinds continue to occur; and rising coal consumption directs attention to the acid rain phenomenon, as well as to the more distant risk of climate modification from carbon dioxide accumulation in the atmosphere. An absence of future energy traumas can no longer be taken for granted. There is a need for insurance of one kind or another to protect against the next shock or surprise. Strategic petroleum stockpiling is about the only protection we have bought so far, though much of the long-term research and development, mostly performed with federal government funding, also falls under that heading (see Chapters 3 and 7 for further discussion).

Reality seven.   "Sound R&D is essential, but there is no simple 'technical fix'." The role of R&D, as detailed in Chapter 4, is, in essence, one of providing energy at a lower cost—both private and social—than would prevail were the R&D not undertaken or not successful. Thus, the accent is on cost. Yet, public imagination is dominated by the search for the *deus ex machina* type of technology that will put us back on easy street. News items and television spots announcing the car that runs on tap water typify this approach. Even less extreme suggestions tend for the most part to be illusions, since they typically ignore or pass lightly over the cost side of the equation. There are lots of ways of generating energy, but only a few are competitive with ways we already know and utilize.

These "realities" set out in the late 1970s seem to have stood up pretty well in a world in which generalizations have an exceedingly brief lifespan. With hindsight, one might have wished to add one more reality.

Reality eight.   "What the energy problem looks like depends on one's place in the scheme of things." Private individual consumers see it one way, commercial users another way, suppliers of oil, gas, coal, and electricity a third way, and politicians—or to use the more dignified term, decision makers—see it in yet another way. Oil exporters look through one end of the telescope, oil-poor consumers in Western Europe, Japan, and the bulk of the developing countries look through the other. Even exporting countries do not share a single perception: small, thinly settled, oil-rich nations like Saudi Arabia or Kuwait will see the problem and the policies to pursue, quite differently from a large, densely populated and less oil-rich exporter like Nigeria or Indonesia. Unless one bears in mind that "where one stands depends on where one sits," the debate over both the roots of the energy problem and its management will often seem more capricious and disjointed than they really are.

A special instance of this reality merits mention. As energy expenditures form a larger share of a household with low income than of one with high income, an increase in energy costs that exceeds the rise in the general price level, disproportionately hurts the poorer segments of the population. Unless countermeasures are taken, rising energy costs will aggravate the consequences of income disparity. Such countermeasures may be supplements to income, assistance in paying for energy, help in conserving energy such as through weatherization schemes, and others. Regions also are affected with differing severity. For most of the 1970s, for example, the price of heating fuel rose faster than that of other energy sources. People living in areas heavily reliant on heating fuel, such as New England, felt the burden of the energy problem more quickly and more sharply than those living, for example, in the North-west where low-cost hydroelectricity provides much of the energy, or in areas like the Southwest that benefited from low-cost natural gas. These relation-ships, however, are not cast in concrete. Price relationships will change in the 1980s, and different areas of the country will come in for their share of distress from rising costs. In any event, the phenomenon of unequal impact raises the issue of remedies: how much, to whom, by whom?

SIZING UP THE ENERGY PROBLEM: FOUR PERSPECTIVES

If the foregoing are the "realities" of the energy world, what are the factors, or circumstances, that make it so difficult to "solve" the problem in an or-derly way? They are (1) that the players all have different interests; (2) that energy is so pervasive in the modern economy that it gets one badly tangled up in other values people hold and in policies that deal with other objectives of our society; and (3) that the complexities of the energy system itself and its interaction with other forces in the economy, both at home and abroad, make it very hard to demonstrate persuasively that any suggested policy would be successful. These entanglements and complexities are briefly laid out in the subsequent paragraphs in terms of the main actors, but in a sense they pervade much of the entire book. Here we examine four perspectives in some detail—those of the individual consumer, the business user, the suppliers of energy, and the government. The theme is dealt with further in the Epilogue.

The Individual Energy User

The principal complaint of the individual energy user, over the last de-cade, no doubt has been the rising costs of keeping the home warm in winter and cool in summer and of driving the family car. At times, there have been fuel shortages and major inconveniences, though not for long and not every-where. Organizations that endeavor to articulate the consumer's viewpoint have argued for holding prices down and have focused on the high profits of the oil companies as one part of the inequitable outcome of rising prices. With the price of much natural gas to be freed from controls over the next few years, the controversy will obviously not die down. Consumers, however, do not limit their views to matters of price. They would prefer rationing to prices as a useful means of allocating supplies in periods of actual shortages; they greatly favor action to produce better information for buyers about energy efficiency of appliances, home insulation, automobile performance, and the like, and see value in tax incentives to propel people toward more energy-conserving behavior. In general, they exhibit a positive attitude toward a role of government and a somewhat skeptical view of the market. For those or-

Who Pays

On and off in this book we use phrases such as "industry bears the cost" or "government will generally fund this activity." These are convenient first-cut, shorthand expressions. In an exhaustive treatment one would want to go on and try to figure out who it is that *eventually* pays. There is no general answer, except that the first judgment is likely to be wrong. In each case the incidence of costs will be different. In principle, it will depend on the degree to which the first party hit by a cost or charge is in a condition to pass it on to the next in line. That, in turn, will depend on how strongly buyers react to increases in the cost of things they wish to acquire, that is, on the "price elasticity of demand," to give it the technical term.

Suppose a producer is forced by government regulation to incur costs to preserve a given environmental good, say, a clean stream. He will try to incorporate that cost in the price of his product; but at the higher price he may lose sales. Thus he may have to "eat" the cost increment, wholly or partially, depending on the characteristics of demand for his product, in which case stockholders, management, or labor bear the burden. There is no disembodied entity, "the corporation" to do so. In turn, competitors that do not face that particular problem are likely to gain in sales and profit. Some consumers may be better off, others worse. It all boils down to this: (1) a clear stream is not the kind of product that has a price tag reading, "I am worth X dollars"; (2) the costs incurred to keep it clean are dollars and cents reflecting the expenditure of real resources; (3) there will be a dollar-and-cents cost to society; (4) the cost will be offset by an unpriced gain; and (5) the costs will be borne by producers and consumers in proportions that vary with the circumstances. Thus the initial judgment of who pays is just that: initial.

ganizations that claim to represent consumers, that skepticism is sharpened by the income inequities toward which the market is neutral and by the inadequacy of information needed by buyers and sellers to make satisfactory decisions.

The Business User

Business users of energy take a somewhat different view. For many of them, energy represents only a small fraction of the final cost of the product or service they sell. Thus reliability—of the right amount at the right time—takes on greater importance than price, though the individual user is not, of course, indifferent to it. If the cost of energy rises, some can recover it in the price they, in turn, will charge, without losing customers. Most will not be so fortunate. Those for whom the cost matters may have the option of switching from one energy source to another—not in a day or month, but over time, depending on their judgment regarding relative future prices. Being market-oriented, they are generally opposed to government intervention in pricing, but favor stability of policy without which they find it hard to plan investments, marketing strategies, and so forth. On the whole they can adapt reasonably well to the higher cost of energy by economizing on its use.

As Chapter 2 points out, industry's performance in energy conservation has been extremely good; but then it is not only easier for a business than for individual citizens to track its cost and to know what to do about it, including access to capital, but its profits will reflect the results of good energy management. Business thus is deeply engaged in conservation in its day-to-day operations, while by and large it has been less vocal than have individual consumers in putting conservation at the top of the energy agenda. Finally, firms favor the least feasible amount of government intervention (though they naturally tend to be partial to types of "interference" that benefit their particular activity).

The Energy Supplier

There is a widespread impression that energy suppliers have so far had the best of the energy problem. But the picture is not that simple. Some have gained and some have lost. Indeed, the energy-supplying sector is far from being an homogeneous group. Cynics even might suggest that perhaps its most common attribute is the hostility that it has evoked from the public. Whether corporate headquarters or heating fuel dealers, whether gas or electric utilities—suppliers have universally been looked upon as the villain of the piece. Suppliers also share something else: an impatience with governmental regulation and processes, to which they tend to ascribe much of the blame for what goes wrong. Otherwise, their views depend on the particular energy source with which they are associated and their place in the structure of the industry.

Thus, to U.S. coal producers, the jump in oil prices is the best thing to come along since World War II briefly boosted the demand for coal to unprecedented levels. By 1980, the industry had more than recovered from the post–World War II decline in coal demand. It was producing at a rate roughly twice that to which it had fallen in the early 1950s and above any level reached in the past. Domestic demand from utilities which switched from oil and gas to coal swelled the rank of customers, and so did foreign countries that in the past had virtually limited their purchases in the United States to coal not for raising steam but as an input into metal production, above all iron and steel.

To the U.S. coal industry, therefore, the energy problem of the 1970s presented not a problem but an opportunity for expansion. Thus focus was on removing obstacles to growth. And there were—and still are—a good many, connected largely with the difficulties of extracting coal in an environmentally acceptable manner and making it a clean-burning fuel (see Chapter 6 for examples of this). Thus, the coal industry has been generally opposed to tightening environmental regulations, both in the mining and the combustion phases. It has also been critical of price controls on oil and gas, in the belief that uncontrolled prices would more fully highlight the price advantage of coal.

It would be a mistake, however, to think of even a particular energy supplier such as the coal industry as monolithically driving toward sharply defined objectives. It is too heterogeneous a group for that to happen. The principal split is between eastern and western coal, with the Mississippi River the rough dividing line. Eastern coal is generally mined underground, is high in sulfur (which is not good) but also high in heating value (which is very good); it is close to domestic markets and suitable for export as well. Western coal is primarily surface-mined, low in sulfur, low in heating value, and distant from the large utility markets. It benefits, however, from access to efficient surface transportation facilities and has gained customers both in the Great Lakes and the Gulf Coast regions. All in all, there are enough differences to bring about conflicts on specific policy issues.

If the energy problem opened up new opportunities for coal, it was a mixed blessing for the petroleum industry. Domestically, rising prices were welcomed and in some years high profits were almost embarrassing. Drilling activities rose steeply for several years, reflecting the gains to be made by finding oil or accelerating its extraction, even though the windfall profits tax diverted a portion of the oil bonanza to the U.S. Treasury. Internationally, there were mostly upheavals and setbacks. Increasingly ownership and management passed from the companies to existing or newly created governmental entities. Chapter 5 presents the evolution in some detail.

While the oil industry is hardly moribund—drilling activity attained record levels in 1981—the 1970s have, nevertheless, hardened the impression that it is an industry struggling to retain its dynamism. As Exxon's Board Chairman, C. C. Garvin, Jr., remarked in a talk to environmentalists on September 30, 1981, "Indeed, we believe, that in the United States oil use reached its peak several years ago and is now on the way down."

Several factors support this assessment and the industry's perceptions based on them. Although the output of Alaskan oil arrested the decline in total domestic oil production, production in the lower 48 states has continued to decline—from a high of 9.41 million barrels per day in 1970 to 6.95 million barrels per day in 1981. The ending of price controls on crude oil has not caused an increase in output, as some had expected, but has stemmed the decline. Heavy investments in some frontier offshore exploration have yielded poor returns, and efforts to squeeze more oil from each well continue to lag. The most massive new discoveries have been made outside the United States, such as in Mexico and the North Sea, though there are high hopes that, especially with regard to gas, the so-called Overthrust Belt—a geological formation extending from Alaska to Mexico—will turn out to be a rich source.

To make matters worse, some of the oil giants elected to make substantial investments in unrelated ventures. Some were in hard-rock mining companies (ARCO buying Anaconda and SOHIO buying Kennecott, for instance) while Mobil bought the Ward stores and Exxon purchased the Reliance Electric Co. and got into the office equipment business. Coal companies had been acquired during an earlier period. Such ventures have called forth exaggerated reactions. They were quite small in terms of both the companies' assets and their investments in *oil-associated* ventures.

A more methodical view of the issue can be found in a tabulation of annual investments in property, plants, and equipment made between 1974 and 1980 by twenty-six major energy companies that report regularly to the Department of Energy. The data, summarized in the November 1982 issue of the Department's *Monthly Energy Review,* show only a small turn toward nonenergy lines of activity. These are the relevant data:

| | *Percentage of 26 companies'* | |
| | *total investment in* | |
Year	*Chemicals*	*Other nonenergy*
1974	5.7	7.9
1975	6.4	7.7
1976	7.1	8.7
1977	8.3	8.9
1978	8.8	8.8
1979	8.1	9.4
1980	8.1	9.3

Chemicals are, of course, closely associated with the energy business. Therefore, the column of most interest is the second. Here the main jump is between 1975 and 1976, and it is not a big one. Even by 1980, about 82 percent of all investment by the companies was in energy activities. Nonetheless, these nonenergy investments tended to feed the public perception that while temporarily floating in cash, the oil industry's faith in the future of oil was fading. Why else would the cash, derived from the higher prices the industry had called for, not be invested wholly in oil ventures? Altogether, for the U.S. oil industry, the 1970s have been a period of adjustment to a changing role and of a struggle to fight off blame for higher energy prices and diminished security of supply.

Different considerations have been shaping the views of the two energy segments that have long been regulated suppliers—the electric and natural gas industries.

For the *electric utilities,* the 1970s have been years of stress. The "energy crisis" has brought mostly bad news:

- The price of fuels used to produce electricity, especially oil, has risen sharply.

- Passing higher costs on to consumers has been a slow process because of the resistance of state regulating agencies, so that costs have generally outpaced revenues. Yet, because the price of electricity has gone up, consumers have complained bitterly about their rising monthly bills.

- The pace of growth in the consumption of electricity has declined sharply, but utilities were locked into construction programs that were geared to historic growth rates and thus proved excessive. Yet, they were too far along to reverse. This has left the companies with excess generating capacity, a costly situation for both producers and consumers.

- High interest rates and delays in licensing and construction have boosted capital costs of new facilities, while poor financial results have made it harder and more costly to borrow. Because it is especially capital-intensive, nuclear power has suffered disproportionately.

Essentially a conservative lot, electric utilities were bombarded with suggestions of novel approaches of setting rates, advising consumers to be efficient, experimenting with new technologies and with cogeneration (a technique more fully dealt with in Chapter 2), and so on.

These approaches, some enacted into law, abounded in the 1970s, and kept the utilities reeling from an embarrassment of advice. Moreover, the movement toward bigger and bigger generating units had begun to produce problems of its own, when such a new issue was least desirable. Added to this, the difficulties with nuclear energy—most of them small but occurring frequently—often lead to prolonged shutdowns and increasingly bad publicity (at times well-deserved). Thus, it is clear why the decade of the seventies spelled trouble for the electric utilities, a condition not likely to abate in the near future.

"The energy problem" may well be a catalyst that eventually will lead to a radical transformation of the industry, be this deregulation, breakup by function (generating, transmission, distribution), decentralization, or nationalization. What is clear in an otherwise clouded crystal ball is that the utilities face severe problems and that they know it. The energy problem is very real for them, but it poses quite different challenges than it does for coal or oil producers.

The *natural gas industry* shares some of the troubles of the electric utility industry. It, too, has suffered a reduction in its growth rate, for natural gas prices have risen—faster than those of electricity—and the high cost of money has interfered with extension of facilities. It, too, has attracted the wrath of consumers. But in several other ways it is more fortunate. It extracts, transmits, and distributes energy, but, unlike the electric utilities, need not convert gas to a different form. Nor does it face serious problems of size or safety. It is not haunted by a Three Mile Island syndrome.

But there are difficulties. For one, the industry does not speak with a single voice. Unlike electric utility companies, most of which produce (generate), transmit, and distribute, the gas industry is divided into different sectors. Gas producers have one set of problems and objectives; the pipeline companies that transmit the gas over long distances have another; and the utilities that retail the gas to individual users yet a third. For example, in the debate over freeing natural gas prices from governmental control, the producers have long been vocal proponents of rapid action, as they would unquestionably benefit from higher prices at the wellhead. The transmitters would prefer a gradual scheme. With a heavy initial or, in the economist's jargon, "sunk" investment in costly pipelines, their interest lies in large and steady volume moving through them, so that the investment can be spread over a large quantity of products, in this instance, natural gas. Their fear that higher prices will cut consumption and lower the volume of gas transmitted, and thus their revenues, keeps them from favoring "instant decontrol." The distributors are in the middle. While they, too, are concerned that consumption growth may be slowed down, at the same time, their revenues might rise, or at least public service commissions would set rates to hold them harmless. Given their sheltered position as a regulated industry, they are not fervent advocates of decontrol, and surely not of so-called old gas, that is, gas discovered prior to the date of the Natural Gas Policy Act (1978) that sets the rules under which gas is now produced and traded.

In summary, there is no one view that the gas industry can be said to hold, nor need any one view be held for very long. What is desirable will change as conditions change. While closely associated with oil, the gas industry is generally more sanguine about its future. For one, the search for gas has a much shorter history than the search for oil, and pleasant surprises are thus more likely. Also, some largely untapped unconventional sources of gas may begin to enter the supply stream when the price rises high enough, though, as discussed in Chapter 3, there is much uncertainty regarding that prospect. Equipped with a vast transmission and distribution network that is fitted uniquely to gas, the industry is greatly interested in novel sources of gaseous energy, either more costly naturally occurring ones, or gas obtained by converting coal or other primary sources. The danger that these additions may be held back by a cessation of the increase or a sustained decline in oil prices now constitutes a major issue for the natural gas industry.

The Government

Finally, there are the policymakers and regulators, or government. In thinking about government, it is useful to bear in mind that government is not itself homogeneous, though we tend to think of it as a well-defined, uniform entity. Yet, there are political parties, elected officials, bureaucrats—all behaving differently. Even so, there are four broad, general issues that delineate the government viewpoint: (1) how energy prices will affect the general price level, that is, whether they will aggravate inflation; (2) how severely and for how long rising energy costs and the massive, sustained transfer of wealth to oil exporters will depress economic growth; (3) whether trends in energy demand, supply, and prices will create sufficient inequities among different parts of the population to require governmental intervention; and (4) whether the nation is equipped to cope with future supply shocks. Most specific governmental initiatives will be seen to involve one or other of these areas. It is worth noting, however, that as the owner of about one-third of all land in the United States, much of it endowed with energy sources, the federal government also has a lively interest in energy matters as a "landholder." Finally, health and safety concerns are very much in the mind of policymakers.

Of course, government always has had a large role in the energy field. In that sense, the events of the seventies did not prove a wrenching experience. About half of the energy reaching final consumers—namely, electricity and natural gas—does so under governmental regulations. The generation of hydroelectric power, originating on rivers under the control of government, was regulated from its very inception decades ago. Many aspects of coal mining, especially those regarding safety, as well as coal transportation rates have long been regulated. So have oil pipeline tariffs. Environmental regulations affect just about every phase of the energy business and have brought government in through the back door. In short, intervention in one phase or another of the energy business existed long before the 1970s.

Thus the many laws passed by Congress since 1974 widened rather than initiated government intervention. The substance or thrust has changed, both with circumstances and with prevailing ideologies, but, interestingly, not in ways that can be tied directly to the political party in power. Thus, the White House under President Nixon reacted to the 1973 embargo and the ensuing oil price revolution with the notion of "Project Independence," which held out the prospect of eliminating oil imports in fairly short order. The administration of Jimmy Carter, while not taking energy independence literally, nonetheless set as goals heavily increased coal and nuclear power production that would enable the country to greatly reduce its dependence on oil. In addition, President Carter imparted a crusading flavor to energy objectives by terming a national energy effort the "moral equivalent of war." In this spirit he gave substantial momentum to conservation, both in terms of information and institutional structure. He also discouraged those phases of nuclear power that might encourage nuclear weapons proliferation—nuclear fuel reprocessing and the breeder reactor. The return of the Republican party to the White House in 1981 did not signal a return to President Nixon's Project Independence. Rather, it revived a laissez-faire philosophy in which decontrolled prices and economic recovery would enable market forces to bring about both greater efficiency in use and increased output of domestic energy sources. The intended result was to be declining imports of oil and reduced dependence on Middle Eastern exporters.

THE CONFLICT IN VALUES

This brief review of how the energy problem appears to different players in the energy game has focused on distinctions peculiar to energy. There is, however, another way of looking at differences in perception and policy prescriptions that is both more complex and uncertain, but more intriguing. It links energy positions to general philosophical inclinations, to values, or, perhaps more accurately, to value systems.

This alignment may at first glance seem far-fetched. What would thinking about petroleum or coal, about refrigerators or electric blankets, about nuclear waste disposal or unconventional gas resources have to do with values? Actually, a great deal. With only a little simplification or license, one can think of most of us perceiving the world around us and ahead of us as offering either opportunities for expansion or the inevitability of limits.

In the expansionist view, economic growth will continue to both require and permit higher levels of energy use, resources will continue to be extended and discovered, technology and, more generally, human ingenuity will find ways to overcome obstacles and to expand our options.

In the limitationist view, economic growth will slow down or cease—indeed, it *should* slow down to preserve both resources and environmental values; technology is a questionable blessing, having brought us a lot of gadgetry as well as the nuclear bomb, and since the real values in life are not material, economic growth is a secondary consideration. Besides, continuing improvements in efficiency of energy use will suffice to depress demand to tolerable levels. In Chapter 2 we show how these beliefs have shaped different people's estimates of future energy "needs."

This notion does not mean that people go around wearing badges reading, "I am an expansionist" or "I am a limitationist." But it does suggest that a great many beliefs come in "packages" and add up to one or the other view of life. Energy is a tailor-made area in which to play out the components of these views.

Why? First, depletion of nonrenewable resources is a fact of life. Concern has flowed and ebbed. But rapidly increasing inroads in the two post–World War II decades produced the "limits to growth" syndrome and philosophy—stated in its starkest form in the 1972 best-seller of that name—which meshes neatly with one of the views described above. From it, but nurtured by various other considerations, flows the denunciation of economic growth as such, and of energy as the major engine of growth, that is held responsible for congestion, noise, litter, waste, and worse. Both economic growth and energy exploitation are held accountable for a "careless technology" that rides roughshod over human values. Thus environmentalism and stress on conservation rather than increasing output—that is, demand rather than supply—are emphasized. Next, it is a fact that energy and the major energy-using artifacts are produced largely by big corporations. Not surprisingly, persons who believe that "small is beautiful" or those who see great merit in decentralization will tend to be hostile to the energy industry. This is not a new phenomenon but one based on an enduring U.S. tradition in which the corporate "robber barons"—be these coal companies, oil trusts, or electric utilities—have been popular targets of attack. If these views characterize one school of thought, their mirror image defines the opposite side: economic growth is the precondition for a better life for all—the economy depends crucially on energy; environmental problems need to be managed but in a confrontation must yield; technology is the great force for progress and needs continuing care; small may be beautiful, but is also inefficient, and so on.

Some may see this as "reaching," but by and large these kinds of issues tend to align people quite neatly in opposing camps. Among other things, these groups hold contrasting views on present and future energy questions just as they are likely to hold contrasting views on other contemporary issues such as abortion, women's rights, nuclear proliferation, developing countries, or public lands, to name just a few. Particularly in the case of energy matters, they will differ on the role of market forces versus government intervention or direct provision of services by government. The fact that energy views tend to be an integral part of these packages of ideas and values has increased the difficulty of developing a national energy policy, since arguments focusing on energy alone are unlikely to change anyone's view on any given facet of energy policy.

It is one thing to speculate broadly about the complexities of the energy problem and to attempt to get a grip on its major facets. It is quite another to form a clear picture of the trends that have developed during the past decade. Without the coordinates provided by a minimum array of data, what follows in the body of the book lacks a dimension of reality. Thus, the succeeding paragraphs provide a quick sketch of where we have been and where we seem to be going.

WHERE MATTERS STAND

Whether one considers it very courageous or very foolish to prognosticate where the United States or the world is heading in its use of energy, one must anticipate a continuing flood of projections and forecasts. The best advice to the consumer of such undertakings is to view them with caution. A feel for the difficulties of looking ahead can be conveyed by a quick look at the history of the past decade. The events of the first few years since the oil shock, say, 1973 to 1978, are easily summarized: the oil embargo; the initial quadrupling of the price of OPEC oil; the rise in domestic prices of all fuels, reflecting higher oil prices; the perturbations in the international trade and balance-of-payments accounts; the accumulation of huge financial surpluses by the oil exporters; and the occasional emergence of physical shortages, particularly in gasoline and in natural gas. The same half-decade also saw a slowdown in nuclear energy growth, as the demand for electric power began to falter, costs of construction continued to increase, and safety concerns complicated and slowed the entire machinery of licensing, site selection, and construction schedules.

A few benchmark data are worth singling out:

1. After the initial price boost of early 1974, energy prices as a whole in the United States rose at about the same rate as the general price level (Figure 1-2). Put differently, they showed only a modest increase in con-

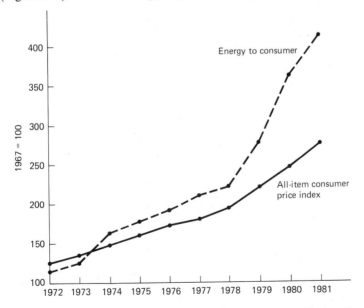

Figure 1-2 The price of energy to consumers relative to general inflation, from 1972 to 1981. After the initial price boost of energy in 1974, the consumer price index for energy rose more or less in step with inflation until 1978. Since then, energy prices have far outpaced the rate of general inflation.

stant dollars until 1978. For regulated energy sources, the increase was initially less steep but at the same time more steady.

2. Efficiency in the use of energy rose steadily, as expressed in the declining ratio of total energy consumption to the GNP (Figure 1-3).

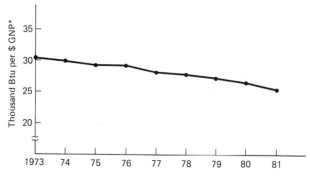

*GNP in constant (1981) dollar

Figure 1-3   The rising efficiency of U.S. energy use. Each year from 1973 to 1981 it took less energy (fewer Btus) to produce a dollar's worth of goods and services (GNP) in the United States. Producing the 1981 GNP at the 1973 energy efficiency would have taken about 16 quads more than were actually consumed.

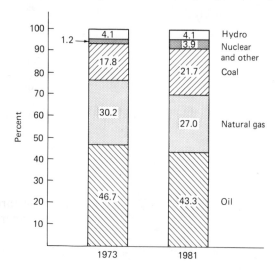

Figure 1-4   The declining importance of oil and natural gas as energy sources, from 1973 to 1981. During the decade after the OPEC oil embargo the United States became a little less dependent on oil and gas and a bit more dependent on coal and nuclear power. Wood is not included in this presentation, but in the 1980s it provided about as much energy as nuclear plants, primarily for the wood-associated industries [Data from U.S. Department of Energy, *Monthly Energy Review,* various issues].

3. The hoped-for expansion of coal's share as an energy source occurred, but less rapidly than government had expected (Figure 1-4).

4. U.S. oil imports continued their upward march until 1977, when they accounted for 47 percent of total U.S. oil consumption (Figure 1-5).

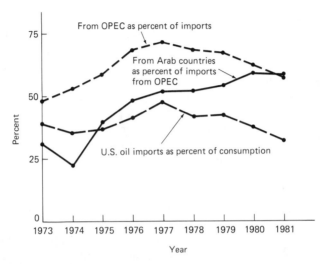

Figure 1-5 The rise and decline of U.S. oil imports, from 1973 to 1981. Net oil imports rose to a peak of nearly 50 percent of U.S. consumption in 1977, and declined to less than 34 percent in 1981. The OPEC share in imports rose to more than 70 percent in 1977, and declined to 56 percent in 1981, still above the 1973 share. The share of Arab members of OPEC rose to a high of nearly 50 percent in 1980, almost twice the 1973 level and declined slightly in 1981. [Based on Hans H. Landsberg, "Relaxed Energy Outlook Masks Continuing Uncertainties," *Science,* vol. 218, no. 4576 (December 3, 1982) figure 1.]

5. OPEC oil continued to rise as a share of U.S. oil imports through 1977; within OPEC, the share of the politically most volatile component—oil from its Arab members—increased steeply (Figure 1-5).

6. The rapid rate of growth in nuclear power generation reached a peak in 1978.

7. More fuel-efficient, domestically produced automobiles and high gasoline prices that increased the demand for imports of even more fuel-efficient foreign cars, which in turn pushed domestic producers to exceed government-mandated efficiency standards, caused the fuel efficiency of the U.S. automobile fleet to recover from its all-time low reached in 1973.

The late 1970s and early 1980s have seen some new developments. In the wake of the Iranian revolution, OPEC doubled its oil price. As a result, oil prices to consumers in the United States jumped a second time, though they leveled off and began to decline in early 1981. The combination of the increases, the sluggish performance of the economy, and the bottoming out of the slide in domestic oil production, caused (1) a substantial reduction in oil demand beginning in 1980; (2) an even more drastic decline in oil imports; and (3) a strong continuation of the reversal, begun in 1978, of the adverse oil import trends noted under point 5 above. Net oil imports as a share of consumption fell from 46.5 percent in 1977 to 27.8 percent in 1982, and the share of OPEC oil in imports declined from 70.3 percent in 1977 to 42.9 percent in 1982. Also, for the first time, the share of Arab oil in OPEC shipments to the United States declined, from a high of 59.3 percent in 1980 to 40.8 percent in 1982. Finally, by 1982, with coal exports expanding, the volume of net U.S. imports of all energy sources had dropped more than 50 percent from the high point reached in 1977. These are all welcome developments. What has not changed are the depressed condition of the nuclear power segment and the slower than hoped for growth of coal's share in aggregate energy consumption.

On the whole then, the years from 1973 to 1982 were not a period of unrelieved disaster, either in terms of real energy prices or of consumption trends, though some patterns proved to be resistant to rapid change. Specifically, the beginning of the second decade since the Arab oil embargo of 1973 appears to herald some good news—energy, especially oil consumption, down; oil imports lower; OPEC dependency down; energy production steady; and real energy prices down or rising only moderately. In different ways, these developments all eased earlier constraints and pressures, but unfortunately there was a tradeoff. The "good" news was paid for partly by worldwide general economic distress, high unemployment, and low economic growth (see box on page 24).

In a much longer historical perspective, one is tempted to judge that some of the trends in world oil that set the stage for the 1973 oil crisis, such as the rapid rise in worldwide energy consumption and the growing share of oil in that total, are beginning to undergo permanent modification. It was these twin forces that created the tight market in the early seventies. That factor, in turn, enabled oil producers to boost prices and, what is more important, to maintain them at a higher level. A world built around low-cost petroleum, both as a fact and an expectation, had no quick way of getting off the oil kick.

Unfortunately, there still is no reserve oil output capacity in the United States, a condition that contributed crucially to the success of the oil embargo and price increases in 1973-74. Reserve capacity now lies with OPEC, and lots of it: some 13 million barrels per day in late 1982, or 50 percent more than annual U.S. oil production. But as long as OPEC's members abstain from tapping it in an attempt to increase sales and revenues, the world oil output and price structure is unlikely to undergo radical change. Pressures for increased sales, even at reduced prices, will at times prove too strong for some of the less-affluent OPEC countries to resist, especially as they watch non-OPEC producers steadily increase their share of world markets. In that event we might see a highly turbulent oil and energy market. In the short run, this would spell a welcome period of higher real incomes for oil-importing countries, accompanied by faster economic growth. In the longer run, such a development would postpone but not avoid the switch away from oil to other energy sources. Renewed adjustment problems, if not crises, would arise if oil prices once again were to start an upward climb, a development that appears inevitable.

As we have indicated above, continued growth in coal consumption should occur both here and in many parts of the world. This outlook rests mostly on the comparatively low cost and the certainty of long-term availability of coal, given both the very large resources and the diversity of suppliers. But coal is a bulky fuel with undesirable side effects, and in the long run consumers will want, or may be pressed, to reduce its use. Coal's dilemma is that its environmentally adverse impact, substantially exceeding that of oil and natural gas, can be greatly diminished, but the cost of control in turn diminishes coal's economic attractiveness. Moreover, as is detailed in Chapter 6, the atmospheric accumulation of carbon dioxide, to which coal is a major contributor and whose precise role in altering the global climate is as yet poorly understood, is not at this time judged amenable to correction. Depending on the findings of future research, expansion in coal use may have to end and steps may have to be taken to diminish coal burning altogether on a global scale.

What, then, is waiting in the wings to take the stage when affordably priced oil and natural gas thin out and the use of coal is countermanded? A few years ago, most people would have had an easy answer—indeed, several easy ones. Some would have pointed to the vast oil shale resources of the Rocky Mountains, totaling many times the country's crude oil resources. Others would have named unconventional gas formations of various origins, located in different parts of the country. Still others would have nominated nuclear energy, especially the breeder, citing the well-known fact that it produces more fissionable material than it consumes. Many would have listed a variety of living sources, lumped together as biomass—trees and crops which can be converted to fuel as well as organic waste material. Last, but not least, solar energy would have had many supporters, both in its direct form as a source of radiated heat and in such guises as the wind, ocean waves and temperature differentials, and geothermal sources deep in the earth. With so many alternatives, one might ask, What is the problem? Who needs coal and oil?

Ten years have passed since the benchmark year of 1973, and we are probably all a little wiser now. We have learned that each of these sources and technologies is capable of providing commercially usable energy. In addition,

"Energycentricity"

Preoccupation with energy problems can easily lead one to a loss of perspective, that is, to forget that what is "good" in energy terms is not necessarily good for the economy or for human welfare. For example, there was much rejoicing when U.S. energy consumption in 1981 turned out to be no larger than it had been in 1973, the year the "energy crisis" broke. There was similar satisfaction when oil prices were allowed to rise because they would curtail consumption, both through spurring efficiency and through sheer inability to sustain previous levels of purchases. Good news in a sense, but at a price. It would be hard to deny that we would all be better off if the economy could sustain *lower* energy costs and *greater* energy consumption (provided that environmental effects associated with energy use could be adequately handled). With lower energy costs, real resources—labor and capital—could be employed elsewhere to produce goods and services other than energy which, as we point out, is not wanted for its own sake anyway. Slow economic growth is not a blessing to most people. It reduces available output, reduces the options people have in exercising their demand, exacerbates social strife, and leaves fewer resources to maintain the environment, both natural and man-made. Thus when we welcome measures that reduce energy demand, we should applaud with moderation, recognizing that we are welcoming a second-best, "given the regrettable circumstances," as it were. Increased use of inanimate energy has been mankind's means of escape from drudgery. Neither higher costs nor decreased use are in themselves good news. Only a bad case of "energycentricity" can make one think they are.

we have learned above all that each is afflicted with its own set of problems, some highly publicized and others known only to those who have made a study of them. And then there are cost considerations.

To quickly run down the list, nuclear energy is inescapably handicapped by its associated radiation. Efforts to control it so far have been highly successful, but a major accident can put at risk large areas and numbers of people; in addition, it lends itself to weapons manufacture and thus can help spread possession of "the bomb." Also, management of spent waste from reactor operations is still an unsettled issue. Solar energy, beyond adapting our building patterns to its use, which will take time, has turned out not to be as inexpensive as an earlier focus on the abundance and costlessness of sunshine had led many to believe. Capturing sunlight requires materials, structures, maintenance, and, in most locations, backup energy sources or short-term storage facilities. In short, these supplements or alternatives have so far proved to be technically available but are neither cheap nor free of problems.

Continuing the list, as has been true for over fifty years, oil from shale has remained above the competitive cost threshold, and so have most unconventional sources of gas. They tend to look promising in textbooks and surveys—and at times to investors—but have stubbornly failed to make a significant contribution to energy in the real world. They may do so—and probably will—some day, but repeated disappointments in the past have turned many believers into skeptics.

—Biomass energy prospects round out the picture. The United States has passed through one phase: an experiment with grains. It has been a sobering experience in which a few advocates ran away with the idea for quite a stretch, managed to capture large subsidies, and thus were able to put a product, gasohol, on the market. Gradually, excitement ebbed, as studies showed that without subsidies, the economics of gasohol were not promising; that in the long run the competition for land would, in most parts of the world, place food in conflict with energy; and that the dimensions of liquid fuel demand were such that crop-based fuels could never hope to do more than make a dent in meeting it. Now interest has shifted to wood, and thus to trees, as a source of biomass. Here the verdict is not yet in, but it is likely that in some parts of the world where there are fertile stretches of land not used for agriculture, trees grown specifically to provide fuels could become a viable energy source. Even so, such potential production is unlikely to meet more than a small share of energy uses on a global scale.

To sum up, we are far from making a transition to new sources of energy. In a nutshell, the problem consists of making the most of oil, gas, coal, and whatever nuclear reactors the world decides it can live with, while finding better ways of turning a variety of physically suitable forms of energy into an economically affordable supply. The themes we have touched on in this chapter will be further explored in the chapters that follow.

CHAPTER 2

How Do We Use Energy?

Energy is called on to facilitate so many different tasks in our lives under so many different circumstances that the answer to the question posed by the title of this chapter begins, unavoidably, with the phrase, "It all depends."

It depends on what, specifically, we want energy to do: coke ovens in an iron-and-steel complex and light bulbs in our homes both require energy to function, but they use energy in entirely different ways. It also depends on economics—the cost of energy today, expectations about future prices, the ease of switching to materials and equipment which use less energy. It depends on geography: areas enjoying lower energy prices will tend to consume more of the stuff. And it depends on behavior: certain "energy-intensive" aspects of our life-style—the one-person-per-car commuting pattern, the detached home—set during a period of cheap energy may have become so ingrained and may be so highly valued that they will not be casually given up merely because energy prices have risen. For all these reasons, sweeping characterizations of "a nation's" energy use can be highly misleading. That use is intricately tied to the way society is organized and how the economy operates. (The box on page 27 expands on this point a bit more.) It is a complex story, but one which anyone concerned with energy requirements and efficiency will want to know more about. This chapter tries to unravel the principal strands of that story—the patterns of use and the problems of energy conservation and efficiency.

Explaining Differences in the Degree of Energy Use

As a chemical manufacturer, the proportion of your total production cost that is likely to go for energy can easily be ten to twenty times higher than it would be in the clothing or food-processing industry. Paper, plastics, and basic metals are also highly "energy-intensive" industries, though generally not to the extent that chemicals are. Where energy-intensive activities dominate a region's economy, as in the case of the petrochemical complexes on the U.S. Gulf Coast, that fact is certain to show up as a much higher ratio of energy use to measures of overall economic activity than elsewhere, such as in the New England states.

What is true of regions within a country applies, to some extent, to intercountry differences as well. The fact that Canada's ratio of energy use to gross domestic product (GDP) exceeds that of the United States reflects the former country's disproportionate dependence on energy-intensive industries.

But other differences in energy–GDP ratios among countries arise from a complex of additional reasons. For example, shorter distances and greater population densities have permitted more energy-saving transportation systems and housing patterns in Western Europe and Japan than in the United States. Moreover, a historical tradition of higher energy prices—particularly because of steeply taxed motor fuel—has in those countries long spurred the conservation practices now beginning to be embraced in North America. The following listing shows comparative energy–GDP ratios for 1980. They are expressed in index numbers, with the United States designated as 100.

Canada	108
United States	100
West Germany	66
France	52
The Netherlands	78
United Kingdom	69
Italy	50
Japan	48

TRANSFORMATION OF ENERGY: FROM POINT
OF PRODUCTION TO POINT OF USE

It is a trite observation that energy resources provide no inherent utility; there is nothing pleasurable about feeling a lump of coal or contemplating a canister of butane gas. Rather, energy is prized for sustaining and enriching our material well-being. Still, it is not entirely pointless to reflect on that simplistic proposition, for it points up some fundamentals about energy use in a modern economy:

- Since energy is regarded highly only for its contribution to providing the products society desires, there is an incentive to meet those goals with less, rather than more, energy—for example, more fuel-efficient cars that preserve comfort and mobility. It is maximizing the "amount of welfare" we can wring from a given amount of energy—not maximizing energy use—that counts.

- Cost and technological uncertainty, however, limit our flexibility in the selection of which kind of energy serves which needs. We may prefer to dispense with oil imports, but probably not at the expense of replacing them with the much more costly, and technically problematic, extraction of liquids from shale or coal. Solar photovoltaic cells will no doubt achieve commercial standing and reliability, but, for now, are no homeowner's substitute for much cheaper, conventional electricity.

- Moreover, while as a physical abstraction, *all* energy denotes the capacity for work, as a practical matter, not all types of energy can do the same jobs. Hydroelectricity cannot energize our automobile fleet, still dependent almost exclusively on petroleum; and carbon-rich coal, rather than liquid fuel, is needed in blast furnaces. We must be concerned, in other words, with access to energy of the kind specifically required to carry on the many complex tasks of a modern industrial society.

For energy to be available in a form appropriate to the task in question, numerous transformations must occur between the raw material ("primary") stage of production and its deployment at final point of use. This chain may involve a fundamental alteration in the physical attributes of the resource at each stage. Coal, for example, is transformed into radiant energy for incandescent lighting as its combustion produces steam, which spins a turbine which—by means of a rotating generator—produces electricity, whose current is forced through a high-resistance filament to give off the vivid glow of an electric lamp.

Energy stage	Form of the product
Chemical	Coal
↓	↓
Thermal	Steam
↓	↓
Kinetic	Rotating turbine
↓	↓
Electrical	Electricity
↓	↓
Radiant	Lighting

In other instances—for example, the conversion of crude oil into gasoline—the properties of the resource remain largely intact, although the process of refining and blending may change important molecular and combustion characteristics of the product. Moreover, the final stage in the use of motor fuel also involves a basic change: the energy content of a chemical fuel being transformed via combustion into the mechanical—or kinetic—energy of the drive shaft. Natural gas undergoes the least change of all, as it moves from wellhead to homes, industry, and farms.

The foregoing suggests there are different ways to measure the nation's energy consumption—at the primary stage (how much coal or oil is supplied), or at the end-use stage (how much gasoline or electricity is used), with the latter capable of being arrayed by major functions in specific sectors of the economy (how much natural gas for crop drying in agriculture).

A common statistical representation is the distribution, by source, of primary energy consumption. This appears in Table 2-1 for the year 1981. The so-called fossil fuels—oil, gas, and coal—dominate the picture, with 90 percent; hydroelectricity and nuclear power share the balance in roughly equal proportions, although, over time, the hydroelectric share has tended to fall while the nuclear fraction assumed an upward course. The estimated 2 to 3 percent contribution from geothermal and "renewables"—the latter virtually all wood, since solar is still of little quantitative importance—rounds out the picture. All told, nearly one-third of primary energy resources used in any single year are devoted to the production of electricity.

The 1981 estimate of U.S. energy consumption, 76 quadrillion British thermal units (Btu), represents the combined contribution of the various energy sources deployed in the United States. (One quadrillion Btu are frequently denoted as one "quad.") The familiar measures of energy use are tons of coal, cubic feet of natural gas, or barrels of oil. But the conversion of different physical units into a common denominator—a measurement of energy based on capacity for work, namely, the Btu—yields a quantitative representation of energy consumption in the aggregate. (Other common-denominator units, such as equivalent barrels of oil a day, could be used as well and are.)

TABLE 2-1.   U.S. Energy Consumption by Source, 1981

Source	Original physical units	Energy units (quads)[a]	Percentage of total quads
Coal	729 million tons	16.0	21.2
Natural gas	19.3 trillion cubic feet	19.8	26.1
Petroleum	5840 million barrels[b]	32.0	42.2
Hydroelectric power	287 billion kWh	3.0	3.9
Nuclear electric power	273 billion kWh	2.9	3.8
Other[c]	—	2.1	2.8
Total		75.8	100.0

Note: Consumption is measured here at the resource-input or "primary" stage, which is the most commonly used yardstick. In the case of hydroelectric power, the figure in quads shows the quantity of fossil fuels (coal, natural gas, or petroleum) which would have been required to produce the amount shown in kilowatt-hours (kWh).

Source: U.S. Department of Energy, *Monthly Energy Review,* June 1982 (except "other" line, which is a rough estimate). The physical units were converted from British thermal units using the table of equivalents shown in the *Monthly Energy Review.*

[a] "Quads is a shorthand expression for quadrillion (or 10^{15}) British thermal units (Btu).

[b] Equal to 16 million barrels per day. "Barrels per day" is the more common designation for petroleum production or consumption.

[c] Rough estimate for geothermal and biomass (principally wood) products. Since it covers a mix of items, no single original physical unit applies.

It helps to clarify one measurement quirk which shows up in Table 2-1. Electricity generated from, say, coal is not treated as consumption of a primary energy source; it is the *coal* going into the production of electricity that constitutes the primary energy source. When we come to waterpower or nuclear energy, however, measurement tradition treats the very generation of electricity from those sources as primary. And so the energy statisticians assign a primary equivalent value (shown in the third column of Table 2-1) to the hydro and nuclear components of the nation's energy consumption.

A common unit of measure, such as the Btu, makes sense because in many applications and through various processes of conversion, different energy resources and energy forms *can be substituted for one another.* (It is precisely because of such alternative possibilities that economists and government officials worry about the presence of competitive or anticompetitive forces among firms in the energy industry. This subject is taken up in Chapter 5.) Electricity can be produced by any fossil fuel, nuclear, hydropower, and a variety of other sources. (Figure 2-1 shows the resources used to generate electricity in the United States in 1981.) Oil, gas, and electricity continue to compete, and in the past competed actively, in residential and commercial heating of both space and water; solar systems are a more recent option. A variety of industrial processes involving steam or direct heat frequently permits choosing among coal, oil, or gas. The final choice in many of these in-

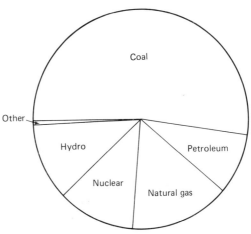

Figure 2-1 Resources used to generate electricity in the United States, 1981. Electricity can be produced from a number of alternative energy sources. The choice is influenced by supply adequacy, geography, technological and environmental considerations, and government policies—all of which affect the comparative cost of the different possibilities. [Data from U.S. Department of Energy, *Monthly Energy Review,* May 1982.]

stances is heavily influenced by cost—in the coal-producing states of Appalachia, electricity production is most economically based on that fuel—but related factors, such as technological uncertainties, pollution-control requirements, and government policy measures, obviously enter into decisions taken.

In numerous circumstances, of course, the substitution of one energy form for another is limited. How could Britain and certain American cities have illuminated their streets in the 1860s without gas manufactured from coal? These days, the automotive market is essentially captive to petroleum—though, given time, willingness to pay, and accelerated technological progress, we could at some point get the needed liquid fuel from coal or shale; or achieve success in development of an electric vehicle. Lighting, mechanical drive, and certain electroprocess industries (for example, aluminum smelting) are keyed to electricity. Moreover, in a host of critical applications (representing over 10 percent of energy demand by industry) oil and natural gas are not used as fuels; instead, in the form of "petrochemical feedstocks," they are combined with other materials in the manufacture of products such as synthetic fibers, plastics, pharmaceuticals, fertilizers, and lubricants.

SPOTLIGHTING PARTICULAR ENERGY USES

Figures 2-2 and 2-3 present a handy summary of the consuming sectors, activities, and energy forms which dominate energy use in the United States. A couple of things need to be said about the presentation. The breakdowns shown are very approximate. There is no systematic collection and validation process, either in or outside of government, for such a compilation of data, so one must make do with infrequent and rough estimating efforts. The labels of a

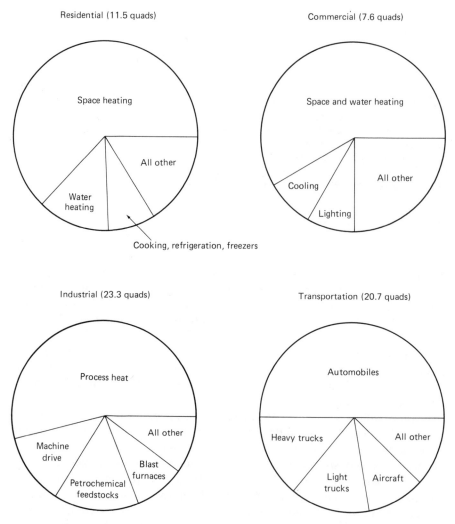

Figure 2-2   Pattern of energy consumption, using sector and function, 1978. Note how energy use in each of the four consuming sectors is dominated by one or two specific functions. The sum of the sectoral consumption figures shown in parentheses (63.1 quads) falls short of nationwide primary energy consumption (79.7 quads) by the amount (16.6 quads) of "conversion losses"—principally heat lost in converting fuels to electricity. [Based on data from Table A-1.]

few items in Figure 2–2 may not be part of the reader's everyday vocabulary. "Process heat" covers both direct firing (such as smelting) as well as indirect heat applications involving, principally, the use of steam in chemicals manufacture, petroleum refining, papermaking, and numerous other activities. In the manufacture of paper, for example, steam is required in the chemical proc-

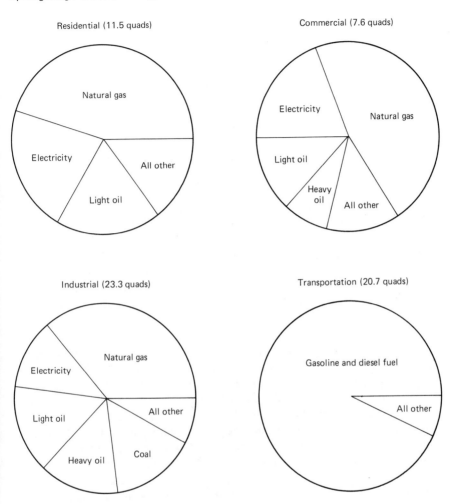

Residential (11.5 quads)

Natural gas

Electricity

All other

Light oil

Commercial (7.6 quads)

Electricity

Natural gas

Light oil

Heavy oil

All other

Industrial (23.3 quads)

Natural gas

Electricity

Light oil

All other

Heavy oil

Coal

Transportation (20.7 quads)

Gasoline and diesel fuel

All other

Figure 2-3 Pattern of energy consumption, using sector and form, 1978. Except for electrified railways, the transportation market is largely "captive" to liquid fuels. As the figure shows, other sectors have more latitude about their preferred energy form. [Data from sources shown in Table A-1.]

ess by which wood is converted into an acceptable form of pulp. (Steam probably accounts for two-thirds of the industrial process heat category, as shown in Figure 2-2.) Industrial use of electricity is dominated, as the figure implies, by mechanical drive—in other words, motors. This is hardly surprising. In addition, electric power in industry is used for electrolysis in chlorine production, aluminum smelting, and other manufacturing steps requiring that process for chemical separation of materials. (Readers interested in more detail than that shown in Figures 2-2 and 2-3 can refer to Table A-1 at the end of this chapter.)

Its approximate nature notwithstanding, Figure 2–2 explains a good deal about contemporary U.S. energy use. Perhaps the most notable thing it tells is how energy use in each of the four consuming sectors is dominated by one or two specific functions:

- Space heating alone accounts for nearly two-thirds of household energy use. Water heating raises the share to more than 75 percent.
- The same two functions represent approximately 60 percent of energy use in the commercial sector, which is defined to include office buildings, stores, and various public facilities, such as schools and hospitals.
- Process heat accounts for well over half of industrial energy consumption.
- Passenger cars absorb half the energy in the transport sector.

All told, this small cluster of uses—comfort and water heating in buildings, industrial process heat, and automobiles—represents 58 percent of U.S. energy use. Contrast that finding with the fact that another group of highly important uses—the air conditioning of buildings and passenger transport by air, bus, and rail—absorbs a mere 5 percent of nationwide energy consumption. It is little wonder that, in terms of national attention, the energy efficiency with which, say, we heat our homes or run our cars gets such prominent play. Substantial savings in just those two categories can mean notable overall energy savings. Of course, even where a particular function signifies only modest energy demand by national standards, that demand can be so important a cost factor in the particular activity that conservation impulses are set in motion. Airline fuel bills, which have become a significant proportion of that industry's costs, have underscored the benefits of the moves—under way even before the energy turmoil of the early 1970s—to modernize the fleet with a new generation of energy-efficient aircraft. This issue—the prospects for conserving energy through its more efficient use—has become a crucial one for individual households, business, and industry.

CONSERVATION

"Conservation" means different things in different situations. In its ethical, equity, or environmental dimensions, conservation has to do with moderation as an inherently worthy pursuit. In the face of possible resource limitations and environmental uncertainties, it is nothing more than a commitment to equal justice under the grudging law of nature.

A student of thermodynamics views energy conservation and efficiency within the bounds of physical laws and processes. The law of the conservation of energy says energy cannot be created or destroyed, only transformed; but such transformation involves inescapable "losses" in the form of unrecapturable heat—for example, because of air resistance and friction in automotive propulsion.

From an engineering perspective, the concern may be with efforts to develop technology and materials boosting the yield of a given amount of energy. Improving the efficiency with which heat is delivered by a residential gas furnace is one such example. Of course, the cost of achieving such an improvement can be fairly critical. An overenthusiastic engineer wholly oblivious to that fact may, in time, be invited to seriously consider a career change.

The last thought alerts us to the fact that conservation must be carefully evaluated as an economic proposition. From that vantage point, which is the one of this book, conserving energy means the most economic application of energy in the production of goods and services or in their use by ultimate consumers. Needless to say, little is gained if all that happens is replacement of energy by something else—materials or labor—that costs more. There has to be some overall economic benefit. Thus, an aluminum manufacturer may be able to *lower overall production costs* by introducing a process that lowers the energy requirement of an ingot produced. A homeowner may succeed in reducing heat losses by adding insulation that costs less than the resultant savings on a year's utility bills. Both cases signify the exploitation of energy-saving opportunities which, if forgone, mean unnecessary expenses. Conservation is not necessarily painless—rear seat passengers in small cars are not as comfortable as they used to be in large cars, but economically it may be less painful.

This dollar-and-cents approach need not conflict with the other perspectives on conservation. If an energy-using process violates physical principles, it is unlikely to be listed in the Yellow Pages. If the technology is intractable, the cost will price it out of the market. If the cost of energy reflects growing scarcity and environmental protection requirements—for example, strip-mine reclamation—thus causing the demand for energy to be less than if those factors were not sufficiently built into its price, then the cause of enlightened resource management will also have been served by an economic yardstick. Of course, government, through its regulatory policy, from time to time prevents prices from signaling changes in supply and demand; it did so in the 1970s, for example, with a cap on certain categories of domestically produced oil in the face of sharply rising world oil prices. But when it resists that price-setting impulse, economics can be a powerful driving force toward energy efficiency.

Consider developments during the period following the 1973–74 Arab oil embargo. As Figure 2-4 shows, certain prices since that milestone have moved steeply higher, in spite of the regulatory policies noted above. (We are talking about "real" prices—that is, prices adjusted for the general rate of inflation.) Notwithstanding periodic intervals of leveling off, the trend between the years 1973 and 1981 is unambiguous. All energy prices rose faster than the general

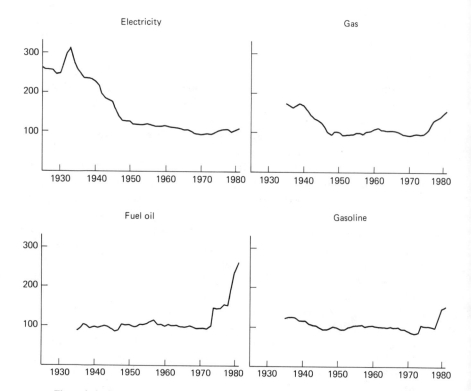

Figure 2-4   Long-term trends in real energy prices to consumers, from 1925 to 1981 (Index numbers 1967 = 100). The rise in energy prices since the early 1970s contrasts with an earlier period of relative price stability or actual decline. Note that in this chart energy prices are expressed in index numbers with a base year of 1967 = 100. The series refers to "real" prices—the movement in energy prices having been adjusted for changes in the overall consumer price index. The series shown are all components of the Bureau of Labor Statistics' (BLS) Consumer Price Index (CPI). The figures used in the chart were deflated by the overall CPI. [Based on data from the Bureau of the Census, *Historical Statistics of the United States: Colonial Times to 1970* (Washington, GPO, 1975); Bureau of Labor Statistics, *Handbook of Labor Statistics* (Washington, GPO, 1975 and 1977); and by direct communication from the BLS.]

price index. For that period, and in rounded terms, average annual rates of real price change (that is, increases beyond those affecting goods and services in general) were as follows: imported crude oil, up 22 percent; gasoline, up 8 percent; residential natural gas, up 11 percent; residential heating oil, up 14 percent; and residential electricity, up 4 percent.

In sharp contrast, real food prices dropped by 0.6 percent annually during this period; clothing prices declined 4 percent yearly; and even medical costs, with their widely recognized upward pressure, rose only by 0.6 percent per year in relative terms. This 1973–81 energy price development, as shown by Figure 2-4, contrasts with a preexisting long-run path of unmistakable price stability, or even decline.

In the wake of rising energy costs, users became much more conscious of the need to conserve, although evidence attesting to that trend is slow in emerging and sometimes circumstantial and anecdotal rather than definitive. Conservation impulses may have lagged at first, then picked up speed. The second big jump in world oil prices (the 1979–80 rise accompanying the Iranian revolution) may have been sufficient to persuade energy consumers that major adaptations were necessary. Even so, a complete response to costlier energy frequently awaits acquisition of a more energy-efficient piece of equipment—a furnace, a car—or structure. Such capital assets may turn over in as few as five or in as many as fifty years. Therefore, conservation may begin primarily with better operating practices—for example, car pooling, improved furnace maintenance, or more conscious attention to thermostat settings. (If residential energy conservation depended on the glacial pace that characterizes replacement of the nation's housing stock, progress would similarly be infinitesimally slow.)

The Industrial Sector

Aggregate energy use in 1982 was about 25 percent below what it had been in 1973; yet industrial production was 7 percent higher. Even before the early seventies, industry—responsive to technological advances and other stimuli—had succeeded in economizing on energy use. But higher energy prices seem to have helped accelerate that trend in recent years. There is, moreover, good reason to expect the momentum of industrial conservation to be maintained for the next several decades, although not without a healthy investment climate, and not if manufacturers dismiss the long-term prospect of upward energy cost pressures.

As one would expect, the experience with energy conservation and efficiency varies among industries and companies. Thus, chemical firms and refineries—along with other energy-intensive manufacturers especially sensitive to rising fuel and power costs—particularly have been motivated to conserve. For example, Union Carbide, which produces plastics, batteries, and industrial gases, reported a 20 percent reduction in energy used per pound of material produced between 1972 and 1979 and set a goal for the mid-1980s based on continuation of that trend. AT&T (and presumably its successor companies) is striving for zero energy growth for the remainder of the century. In the electrolytic reduction of aluminum, a prospective shift from the century-old Hall-Héroult technology to an advanced ALCOA smelting process appears to promise a one-third reduction in electricity requirements. Often the dramatic differences between an industry's average energy-using performance, on the one hand, and practices employed in the newer, most technologically advanced plants, on the other, give one clue of conservation potential. Companies are pursuing energy efficiency not only along technological lines. As illustrated by firms like Dow Chemical and Kaiser Aluminum, they are also developing the organizational and managerial structure and the accountability to ensure economically optimal energy utilization.

A major route toward economizing on energy use in industry is via the process of "cogeneration," so-called because it makes effective use of electricity as well as the steam which the very act of generating that electricity through boilers creates. As noted earlier, the largest part of the item labeled "process heat" in the industry portion of Figure 2-2 consists of steam—a ubiquitous industrial necessity which, instead of being produced as a separate operation, can be captured as a by-product of electricity generation at the same industrial facility. When electricity is produced at remote, central power stations, large amounts of heat are routinely and unavoidably emitted into the atmosphere. Utilities typically burn 3 Btu of fuel to produce 1 Btu of electricity. Cogeneration can increase the efficiency of conventional power generation from this 33 percent factor up to 75 percent—or, conversely, reduce losses from 67 to 25 percent. In recent years, under 5 percent of the nation's electricity-generating capacity has involved cogenerated industrial power. (In 1900 the share was 50 percent so the practice is scarcely novel.) Some important legal issues, as well as some less problematic technical ones, await resolution before this—largely economic—method of capturing waste heat can be exploited on an expanded level. For example, industrial facilities must be assured of a market for excess power and a source of backup power. If, as has been suggested, over 10 percent of nationwide power capacity in the year 2000 could be provided through cogeneration, yearly fuel savings could amount to the equivalent of between 1.5 and 2 million barrels of oil per day.

Incidentally, the principle of cogeneration need not be confined to industrial plants requiring electricity and steam. In numerous European urban centers, waste steam (usually converted into hot water) from central power stations provides distributed heat to nearby stores and apartments. This is called "district heating." If more clustered residential patterns develop in the United States, similar arrangements might begin to take hold here.

Automobiles

The demand for automotive energy is changing dramatically. Throughout the 1950s and 1960s, fuel economy of American cars deteriorated as their size grew and power-draining accessories proliferated. Rapid growth in income and population assured rapidly rising car sales. U.S. gasoline consumption increased from 4.1 to 6.7 million barrels per day between 1960 and 1973—an average annual growth rate of nearly 4 percent. Since then, the trend in gasoline use—while somewhat erratic on a year-to-year basis—has actually been slightly downward, despite more drivers driving more cars. Virtually all forecasters—government, industry, research institutions—expect future motor fuel consumption to be substantially below recent levels.

This striking reversal of the historical record is caused principally by the steady improvement in fuel efficiency. U.S. manufacturers are required by law—and also motivated by consumers' sensitivity to fuel costs—to produce cars averaging 27.5 miles per gallon (mpg) by 1985. As these newer cars become a progressively greater portion of the nation's entire automobile fleet, average fuel economy—not merely *new car* performance—will approach the 1985 target (itself likely to be exceeded). Between 1973 and 1981 this fleetwide average improved from 13.1 to 15.5 mpg, aided, no doubt, by substantial imports of energy-efficient Japanese, German, and other cars. Hearing about someone's 45- or 50- mpg car no longer arouses disbelief. It seems a safe bet that, by the year 2000, American cars on the road will, on average, conform to the 1985 new-car target. The Department of Energy estimated in 1980 that, by the year 2020, cars manufactured *then* will average 50 mpg. Performance can be improved both by a continuing shift toward more fuel-conserving, smaller cars in the automobile mix, as well as by technical improvements— for example, turbocharged diesel engines and vehicle weight reductions. The last does not mean necessarily more cramped interiors; the use of lighter fabricating materials, such as aluminum and plastics, can reduce weight significantly across given size classes. At the same time, it is wrong to pretend that nothing is given up by some of these changes. For example, can small cars ever be made as crash-worthy as large cars?

Such improvements in efficiency relieve the burden of gasoline price increases without forcing the consumer to curb automotive travel: a doubling, say, in both fuel price and vehicle efficiency can leave operating costs for a car and its occupants unchanged, travel undiminished, and yet cut fuel consumption in half. Of course, nationwide gasoline use will not shrink that much; growth in population and in the stock of vehicles rules that out.

But the rate of growth even in total automotive travel seems to be slowing. Federal government studies suggest that the growth in demand for automotive "transportation services"—nationwide vehicle-miles traveled (VMT)—is slowing down and may well be on the way to saturation. Estimated VMT rose by approximately 4 percent a year between 1960 and 1978. For the next two decades, the government projects a sharply reduced growth rate of 1 percent annually. The composite effect of all these factors is that the country's automotive fuel consumption could easily be one-third lower by the end of the century.

Residential Buildings

The potential for economizing on the energy needed to heat our homes is a final example of significant conservation opportunities. In 1978 the U.S. housing stock totaled about 77 million occupied residences—houses, apartments, and mobile homes. As many as 55 million of those units are likely to survive till the end of the century, at which point they may constitute nearly 48 percent of the estimated inventory of 115 million housing units. This simple arithmetic underscores the extent to which sustained progress in residential energy consumption depends strongly on what happens in existing structures as well as improvements made in newly built homes. Much attention must obviously focus on fairly humdrum (though not necessarily unimaginative) steps to improve energy use in *today's* homes—leak-plugging, increased insulation, double-glazing, storm window installation, caulking and weather-stripping, night-setback thermostats, careful shopping around when replacing equipment, and good maintenance practices. Penetration of more novel energy-saving features, such as major architectural design changes and active solar systems, will be governed by the rate of expansion of *new* housing. Stories—not to be dismissed—about amazingly low utility bills in such homes show up more and more in newspaper real estate sections.

That conservation has begun to take hold in space-heating practices is hardly open to doubt. For example, the American Gas Association points out that residential natural gas use has been holding steady in recent years, even as the number of consumers of natural gas heating has expanded. But there is also little doubt that vast potential savings remain to be tapped. The cost of using an additional unit of energy to keep a home heated at a certain level of comfort is estimated by most analysts to greatly exceed the cost of achieving the same goal by a conservation expenditure.

A comparison of these two costs is the crux of the conservation decision. As long as it is cheaper to "buy comfort" by, say, insulating rather than by buying fuel oil or natural gas, conservation is cost-effective. Of course, the payoff from successively greater investments in energy savings becomes less and less attractive—the twentieth inch of attic insulation may cost the same as the sixth but produces little, if any, additional fuel savings—and, therefore, one must be aware of the dollar break-even point as one maneuvers between missing out by underconserving or suffering dollar losses from overconserving. The conservation barrel is no more bottomless than the oil barrel. But there is a long way to go before the bottom of that barrel is reached. The Department of Energy, for example, has estimated that various efficiency improvements can save as much as 40 percent of space-heating energy in existing homes and that the cost of introducing these improvements works out roughly to about $10 to $16 per barrel of oil saved—in other words, substantially below the cost of the energy otherwise needed. In the economist's terminology, the marginal cost of conservation would still fall short of the marginal cost of fuel.

What stands in the way of reaching that potential? There are probably various reasons for the inertia, not the least of which is a lingering unawareness of the economic benefits of increased energy efficiency. Some of the necessary calculations do really require a fair degree of sophistication, as the box on page 42 illustrates. Moreover, it is common to avoid outlays whose reward lies in the future. After all, anything can happen in the meantime (such as moving!). Nevertheless, as time goes on, one can surely count on bread-and-butter instincts to prevail.

The outlook therefore is for future space-heating energy requirements to break sharply with the past. During the fifteen-year period 1960–75, energy use for that purpose increased by one-third. For the remainder of the century, the Department of Energy forecasts an actual decline—about 60 percent of the underlying conservation stemming from modification in the structure ("thermal integrity"), such as insulation; and 40 percent arising from improved efficiency of equipment.

ENERGY AND THE ECONOMY

It is not difficult—and not uncommon—to build a case for a far-reaching *nationwide* trend by means of cleverly selected examples consistent with such a postulated trend. But in the matter of energy conservation, the evidence for the whole economy is not merely inferential: the data strongly suggest a momentum in line with the specific sectoral illustrations which we have given.

The overall performance of the economy is usually indicated by its gross national product (GNP), a measure of goods and services produced or of income generated in the course of that production. When, for an extended period of time, the GNP grows decisively faster than consumption of energy (compared to a preexisting period during which the two measures paralleled each other much more closely), we are entitled to observe that the American economy is using energy more efficiently. As we have already noted in Chapter 1 (and depicted in Figure 1–3), such a trend is precisely what has taken place during the past decade of rising energy prices. Table 2–2 summarizes that development.

TABLE 2-2.   A Comparison of Trends in Energy and the GNP, 1950–82
(average annual percentage rate of change)

	1950–60	1960–73	1973–82
Energy use	2.7	4.1	− 0.6
GNP	3.2	4.2	2.1

Note: Change in energy use is based on quads; and change in the GNP is based on constant (1972) dollars.

Source: U.S. Department of Energy, *Monthly Energy Review,* February 1983; Energy Information Administration, *Annual Report 1980;* and Council of Economic Advisers, *Economic Report of the President,* February 1982 and 1983.

What Goes Into a Conservation-Investment Decision?

Some conservation decisions—say, repairing a dripping hot-water faucet—require little deliberation. Others require a lot. Suppose a household wanted to reduce heating costs and determined that:

- The house consumes, on the average, 125 million Btu of natural gas per heating season which, at the area's prevailing price of $4.50 per million Btu, costs $562.50.
- A $400 investment in improved insulation (assumed to last at least fifteen years) will yield annual natural gas savings of 20 percent—which translates into 25 million Btu or $112.50.

This turns out to be a worthwhile investment. (Indeed, if one had access to the capital, one should consider an even larger outlay, as the Department of Energy analysis cited on page 40 of the text suggests.) For with an assumed interest rate of 10 percent paid on a three-year, $400 home-improvement loan—

- The *unit cost of saving energy* works out to roughly $2.10 per million Btu, significantly below the price of gas at $4.50 per million Btu.
- The simple *payback period* required to recoup the investment is approximately four years, substantially less than its expected lifetime.
- The *benefit–cost ratio,* which is the result of dividing the cumulative dollar value of the energy savings by the cost of bringing those savings about, comes to a positive and healthy 2 to 1.
- The average *annual return on investing* the $400—the homeowner's net profit—works out to about 14 percent over the life of the investment.

An investment producing these benefits and recouped in four years is very respectable. True, there are people who—in economic jargon—"discount the future" heavily, placing much greater value on having money now rather than later. They are frequently impatient with the prospect of even that attractive a payback.

Still other factors figure in the conservation decision. Those who expect future energy prices to rise will, if they turn out to be right, benefit from still greater savings. Depending on one's tax bracket, availability of certain federal or state conservation tax credits and the fact that energy savings obviously cannot be taxed (being a reduction in money that would have had to be spent rather than income received) are further pluses. (Hence the 14 percent return on investment is likely to be understated.)

But some aspects of the decision can be problematical. Will the house, if put on the market, command a price reflecting the value of the conservation investment? What about the quality of the insulation or even possible health effects—for example, indoor air pollution from "overtight" insulation?

Big conservation investments—for example, solar hot-water systems—demand still more careful analysis. Consumers are now able to obtain professional advice and information on these matters, but the whole conservation field has not entirely emerged from its shakedown stage.

Let us be emphatic: the nationwide relationship between energy use and the GNP can fluctuate and change for reasons other than price-induced conservation efforts. Indeed, for several decades following World War I, energy consumption trailed GNP growth even as *falling* real energy prices might have been expected to intensify, rather than to moderate, energy-use growth. The spread of energy-related technologies (factory electrification, farm mechanization, railroad dieselization) appears to have made it possible to squeeze progressively more useful economic output from a given input of energy. Labor productivity—benefiting from abundant amounts of the needed energy—flourished. In part, this development was undoubtedly associated with the availability, in turn, of the newer energy forms. After all, within less than a century, the mainstay fuel of the country had been shifting from wood, to coal, to oil and gas, with electricity assuming steadily greater importance as a secondary energy form. These pervasive technological changes, facilitated by versatile and flexible energy forms that were emerging, seem to have more than counterbalanced whatever tendency there might have been to use energy intensively because of its falling price.

Over time, shifts in the "mix" of national output may also cause the rate of energy growth to be slower than that of GNP: services—for example, educational activities—take less energy to "produce" than industrial goods, and a number of service categories are growing as a share of GNP.

All this having been said, the abruptly induced 1973–82 phenomenon of an actual decline in the level of energy use, even as GNP grew by more than 2 percent yearly, almost surely owes its origin to the new energy-cost era into which we were suddenly propelled. In fact, a 1981 study by the Oak Ridge National Laboratory finds that virtually all of this recent decline in the relationship of energy use to GNP is attributable to conservation efforts undertaken in response to the sharp rise in energy prices.

At this point let us step back and observe that this cheering discovery concerning the impact of higher prices has its grim aspect. Increases in the real cost of any input—whether labor, raw material, or other factor—can never be a favorable development, since its function can be sustained only at the expense of some other productive resource. Thus, the increased energy costs that are the very driving force for conservation—especially in the industrial sector—may compel a rearrangement in the production process that hurts the country's economic productivity and GNP. Nor should we always look at the replacement of energy by labor as a beneficial thing. Perhaps some illustrations will help clarify these points:

- A warehousing operation conducted in an economically efficient way may face new choices between, say, the use of forklift trucks and conveyers, on the one hand, and the use of laborers, on the other, after the relationship between energy costs and wages have changed. The new circumstances may call for hiring a few more workers and dispensing with some of the energy-using equipment. But the economic attractiveness of the operation, while maximized at the new conditions, will be below its preexisting level when energy prices were lower.

- One can imagine a textile firm slowing its shift from natural to synthetic fibers as a consequence of the more costly energy "embodied" in, and responsible for, the higher price of synthetics. The firm may be worse off than before but not as badly off as it would have been had it simply acquiesced to the effect of rising fuel costs. (Passing the effect of higher energy costs forward by raising product prices to the customer does not make such cost increases less real; it only shifts their burden.)

Each of these hypothetical examples, if spun out into its national implications, is consistent with *both* a declining relationship of energy to GNP and a slowdown in GNP growth itself (see Table 2-2). The factors contributing to discouraging performance in productivity and economic growth during the 1970s are difficult to sort out. But the disappointing record in part must be attributable to energy price shocks. It goes without saying that consumers are not spared these jolts. They get the message when the effect of higher energy prices cuts into their real income and forces them to adapt, with least hardship, to the new conditions.

Although society has suffered a permanent loss of income and wealth as a result of the rise in real energy costs, in time, resilience and adaptation should mute these impacts. Other sources of productivity growth can reassert themselves and swamp any remaining negative energy effects. Numerous studies suggest that, in the United States, GNP growth approaching 3 percent per year during the next several decades may materialize in the presence of a vastly lower rate of energy growth—say, one-half to 1 percent (more about this in the next section). As our earlier discussion indicated, conditions are ripe for significant improvements in energy use even without the jolt of further price increases.

Of course, if the easing in oil prices during 1982–83 was thought to herald a prolonged revival of cheap energy, conservation could flop. But such an interpretation of events probably would be both hasty and imprudent.

Our discussion of energy's changing role in the economy has focused on the marketplace as the principal agent of that change. But that is not to dismiss wholly the role of government. Conservation responses, for example, can be strengthened by policy measures that seek to compensate for situations where marketplace adjustments are "sticky." In some circumstances, taxes, standards, and regulations have their place. For example, local ordinances prescribing minimum insulation standards that are clearly in the economic interest of the home buyer may discourage builders from skimping on investments of that kind. Though such skimping may provide a competitive edge by lowering the purchase price of the house, it will do so at the expense of far higher— and more easily obscured—costs spread over the life of the dwelling. Energy efficiency labeling of appliances provides useful guidance to buyers. The principle of providing financial assistance to help low-income families weatherize homes cannot be faulted easily. Energy taxes (which tend to reduce energy use) can help remedy those "market failures" where the price fails to include an allowance for, say, the risks—and potential costs—of international oil-supply disruptions or for environmental damage resulting from energy production or use. The fuel economy standards have given Detroit a useful nudge toward small-car production and efficient engine development, although there is a lively and unsettled debate as to whether that outcome would not have evolved as a strictly market phenomenon. Those believing that it would do so cite the fact that fuel efficiency has been improving at a faster rate than federal law mandates.

There is thus a clear-cut place for a government role that ensures better adjustments by society—including conservation—to energy perturbations. Regulations are not inherently ill-conceived; unregulated markets are not inherently efficient. Yet caution is in order. The principal drawback, and it is not a trivial one, of the compulsory approach to inducing energy efficiency is that it may sometimes impel government lawmakers and regulators to presume and determine technical and economic feasibility rather than permitting such choices to emerge from the dynamics of risk taking and decision making by those frequently in a better position to know and to act: buyers, sellers, manufacturers, investors, and managers. Suppose the government decreed an average new-car fuel-economy standard of 42 mpg by 1990. The target may or may not be economically efficient. But a misguided dictate could impose costs in materials, wages, and capital significantly in excess of the benefits the target is designed to confer. Proliferation of such regulations invites a misallocation of national resources, lessened product diversity, and a deterioration in overall economic performance. This is just one more complex "energy-and-the-economy" interconnection of which we ought not to lose sight.

ENERGY USE IN THE YEARS AHEAD

How much energy will be demanded in the future? It may make more sense to talk around that question rather than to defend a particular figure. During the decade of the 1970s, the art or science—depending on one's view of the matter—of forecasting energy demand approached the dimensions of a national obsession. There were probably two reasons for this development. One was the oil price shock of 1973–74 which, along with its reverberations to other energy sectors, seemed to portend disturbances and discontinuity in the growth and structure of the economy, in technology, and in personal well-being. True, there had been intermittent earlier preoccupation with future energy adequacy and needs, as recalled elsewhere (see Chapters 1 and 3). By and large, however, such concerns did not endure: a condition of plentiful supplies and stable or declining real energy prices had been a fact of life for so long that its presumed permanence made long-range energy assessments unnecessary. In the seventies this was no longer true.

The second reason for the proliferation of future-outlook studies in the energy field goes back to a phenomenon already touched on in Chapter 1. Energy use was becoming a proxy for, on the one hand, the wherewithal for continued economic growth or, on the other, the route toward an ecological disaster. To *expansionists*—flippantly characterized by some as the "dig we must" school—the increased use of energy facilitated a rise in living standards; to the *limitationists,* it simply intensified the growing problems of the environment, safety, and public health.

Of course, these are representations of extremes in the ideological spectrum. On the economic growth side there were some who appreciated that, by exploiting cost-effective conservation opportunities, more sparing energy use would not significantly jeopardize economic growth; while some environmentalists recognized that imaginative policies and careful management could mute the destructive consequences of higher energy production and consumption. (Numerous economists, perceiving the problem not as a matter of ideology but primarily as one of efficiency—enabling market prices to shape energy and environmental adjustments—found it congenial to embrace both of these notions.)

Still, the antithetical postures depicted above are not simply a caricature of positions that came to the fore in the course of the 1970s. Published energy-demand forecasts or "scenarios" diverged conspicuously depending on who made them. For example, a major 1975 study by the Edison Electric Institute (the trade association for investor-owned utilities) foresaw a level of primary energy requirements totaling 167 quads in the year 2000 if reasonable economic expectations were to be fulfilled. By contrast, Amory Lovins—a prominent environmentalist who, in many ways, personified the concern with "limits"—in 1977 argued the reasonableness of a year-2000 target figure of 100 quads, or 40 percent lower. Philosophical positions clearly conditioned analysis. (Even so, Lovins's forecast is likely to turn out to be much more accurate.)

As of 1982–83, a good deal of the sound and fury accompanying the debate over energy in the 1970s has subsided. Why? The record of the last decade has helped shape such a toning-down: the oil price shocks, while severely testing the resilience of the economy—indeed, as we have seen, making it buckle—also demonstrated our clear-cut ability to use energy more efficiently when the cost of failing to do so was seen to be exceedingly burdensome. There has been a healthy process of demystification about energy: when the stakes are high, other productive resources and human ingenuity can serve as a limited, but very important, substitute. Partly as a result of that recognition, there has also been an emerging consensus. Energy-demand forecasts—embodying substantial conservation assumptions—are both far less divergent and much lower (see box on page 48) than they would have been a decade ago. Texaco's expectation about annual gasoline demand over the next two decades (in excess of a one-third decline) is unlikely to be sharply out of line with the views of analysts from the conservation movement. Objective reality has penetrated the mind-set of groups whose views were once very much in contention. The scaling back of future demand estimates has been especially conspicuous on the part of those leaning toward the "expansionist" part of the spectrum. (The other side's projections were already quite low.)

Projections are only as good as the assumptions upon which they are based; and the assumptions up to the early 1970s—that energy would continue to be an easily affordable commodity, and consumed at rapidly rising rates—have been badly shaken. But we should not now conclude from such projections that long-term energy requirements will inevitably continue to have been significantly overestimated. If, as you read this, there happens to be a sharp, boom-induced run up in world oil prices, that episode might—*but need not*—be a harbinger of a permanent state of affairs. (Conversely, depressed price conditions can be a transitory, rather than chronic, phenomenon.) The lesson: try to sift out the ephemeral from what the evidence shows to be the lasting.

It is worth recalling, in that connection, a historical experience with population projections. Numerous demographers in the Depression years of the 1930s were so hypnotized with what they perceived then to be a drastic, enduring fall in the birthrate that they simply could not contemplate a resumption of U.S. population growth. As late as 1947 the Bureau of the Census projected a population total of 164 million for the year 2000—a level reached in 1955!

Although recent energy forecasting has not been crowned with glory, the unavoidable need to plan *today* for circumstances *twenty to thirty years ahead* forces us to continue projecting energy trends, using the best data and analytical tools at hand and working with a range of judgments and assumptions appropriate to the planning decisions which have to be made. Scoffing at the fragility of the forecasting record is not helpful—indeed, it is irresponsible—when, say, as a state utility commissioner, one has to give the go-ahead for a power plant construction program designed to meet energy needs in 2000 and beyond; or as an environmental planner concerned with the greenhouse effect (discussed in Chapter 6), one needs to worry about fossil fuel demand in the twenty-first century.

Perhaps the most important stricture bearing on long-term projections is to allow as much scope as possible for such mid-course corrections as changing economic and political circumstances dictate. New developments, for example, may cast doubt on the degree of nuclear electrification projected just a few years ago; on the pace of solar penetration and acceptance within the marketplace; on the displacement of conventional by synthetic fuels; on the preference of consumers for this or that energy-using product; on the outlook for the economy as a whole and, thereby, the nation's total energy demand.

Energy Projections for the Year 2000: a Contracting Perspective

Successive energy-demand projections for the year 2000 have been progressively reduced. The estimates prepared in 1968 and 1976 came from the U.S. Bureau of Mines (prior to establishment of the Department of Energy). The subsequent projections were all prepared by the Department of Energy.

U.S. Government Energy Demand
Projections for the Year 2000

Year of projection	Quads
1968	169
1976	163
1978	125
1979	108
1980	103
1982	97

A less bullish, long-term economic prognosis is one of the explanations for these steadily reduced estimates. But the recognized potential for, and assumed further exploitation of, energy efficiency opportunities is the principal reason.

For the next several decades at least, the trends and evidence reviewed in this chapter point to only modestly growing demand for fuel and power. The size of the population and size of the economy are expanding far less rapidly than in the past. Improvements in energy efficiency—already spurred by rising energy prices and policy initiatives—are far from fully exploited. At the same time, we cannot afford to ignore the real and enduring contribution energy makes to advancing living standards. The availability of energy to serve that ultimate goal—with efficiency, at affordable cost, with evironmental integrity preserved—is the "big picture" on which we need to keep our eyes trained, even as we remain alert to the course-corrections that a changing world and inevitable surprises force us to make.

BIBLIOGRAPHICAL NOTE

Data on energy consumption in this chapter are derived principally from periodic publications released by the Energy Information Administration (EIA) of the Department of Energy. These include the *Monthly Energy Review* and *Annual Report to Congress*. The latter presents not only statistical data, but also projections, interpretive material, and text discussion. We have also relied on selected energy price series published regularly by the U.S. Bureau of Labor Statistics and on economic data appearing in the annual *Economic Report of the President*.

For those wishing to explore the topics discussed in this chapter more thoroughly, there is an abundant literature. Two books devoted largely to conservation and efficient energy use are John H. Gibbons and William U. Chandler, *Energy: The Conservation Revolution* (New York, Plenum Press, 1981); and Marc H. Ross and Robert H. Williams, *Our Energy: Regaining Control* (New York, McGraw-Hill, 1981). A somewhat technical Department of Energy report, which includes a detailed analysis of energy-conservation issues and potentials, is *Reducing U.S. Oil Vulnerability: Energy Policy for the 1980's* (Office of the Assistant Secretary for Policy and Evaluation, U.S. Department of Energy, November 10, 1980). The interrelationship of energy and economic growth is discussed in Chapters 3 and 6 of Sam H. Schurr and others, *Energy in America's Future: The Choices Before Us* (Baltimore, Md., Johns Hopkins University Press for Resources for the Future, 1979).

APPENDIX

Table A-1 provides a detailed breakdown of energy use by the four major consuming sectors. Figures 2-2 and 2-3 in the text of this chapter are based on this table.

TABLE A-1.   Pattern of Energy Use, by Type of Energy Form, User,
and Function, 1978 (in quads)

Energy form	No. of quads	Sector of use	Specific function	No. of quads
Electricity	2.4	Residential	Space heating	7.3
Natural gas	5.2	= 11.5	Space cooling	0.3
Light oil	2.2		Water heating	1.5
Liquefied				
petroleum gases	1.1		Lighting	0.3
Coal	0.1		Cooking	0.5
Biomass	0.5		Refrigeration	0.3
			Freezers	0.1
			Other	1.1
Electricity	1.7	Commercial	Space and water heating	4.5
Natural gas	2.4	= 7.6	Space cooling	0.6
Light oil	1.3		Lighting	0.6
Heavy oil	1.0		Other	1.8
Asphalt, road oil	1.2			
Electricity	2.7	Industrial[a]	Process heat	12.6
Natural gas	8.5	= 23.3	Machine drive	2.7
Light oil	3.6		Electrolysis	0.5
Heavy oil	3.2		Blast furnace (metal-	
Coal	3.4		lurgical coal)	2.1
Biomass	1.3		Petrochemical feedstocks/	
Other	0.5		nonenergy	3.5
			Other	1.9
Gasoline, diesel	19.2	Transportation	Automobiles	10.4
Heavy oil	1.1	= 20.7	Light trucks	2.9
Pipeline (gas)	0.5		Aircraft	2.1
Electricity	b		Bus and passenger rail	0.2
			Heavy truck	2.9
			Rail freight	0.7
			Marine	1.1
			Pipeline (gas)	0.5

U.S. total = 63.1
Plus: Electricity
Conversion loss (16.6)
= Primary energy
consumption of 79.7

Note: Numbers may not add to totals due to rounding.

Source: U.S. Department of Energy, Energy Information Administration, *1980 Annual Report to Congress,* Vol. 3, *Forecasts* (Washington, D.C., 1981) pp. 124 and 142–143; supplemented by *Report on Building a Sustainable Future* (Washington, D.C., U.S. House Committee on Energy and Commerce, April 1981) pp. 502–505.

[a] Including manufacturing (84 percent of industrial total), mining (8 percent), construction (4 percent), and agriculture (4 percent).

[b] Less than 0.05.

CHAPTER 3

Energy Resources

Human and animal energy, or so-called animate energy, came first. It could accomplish a great many mechanical tasks by lifting, pulling, or pushing. It built the pyramids, dug furrows, and constructed irrigation works. Wind, water, and firewood added inanimate energy, which came increasingly to be applied to more complex and sophisticated uses. Early on, it provided warm dwellings and hot food; then it came to be used for fashioning tools; later still it replaced animal power with mechanically propelled transportation; and finally it became the mainspring of modern industry.

At every stage there was concern over what today we call "adequacy of supply." When fuelwood was valuable—as it is today in many developing countries of Africa, Asia, and Latin America—those who owned a stand of trees guarded it carefully against intruders. Time horizons were extremely short, people lived a day-to-day existence with tomorrow's supplies subject to capture, disappointment, and shortages. To be sure, like wind or flowing rivers that moved ships, fuelwood was a renewable resource. But renewable resources were not reliable. The wind could stop blowing, rivers could become too shallow or too turbulent for navigation, and fuelwood might not be renewable in sufficient volume at a given location. It is useful to recall these limitations since many now look to renewable sources of energy as the great panacea. While renewable resources gave us our start, renewables are also what we left behind.

It was coal that broke the uneasy dependence on poorly controllable natural energy sources. Coal could be mined in quantities demanded and could be stored or shipped to where it was needed. Thus constraints of time and place were eased. Moreover, mechanical power, supplied by carefully regulated coal-based steam, opened new frontiers, as did improved techniques for extraction and smelting of metallic ores used in the manufacture of sturdier equipment.

All this has been well documented in other sources. What is of interest here is that with the realization that large stores of energy existed which could be produced according to some organized plan rather than on a catch-as-catch-can basis, the inevitable question arose, How much is there, and how long will it last? Serious concern with long-term supply began with coal. Although coal was mined in Europe as early as the 1600s and was used by North American Indians at an even earlier time, the issue of long-term supply was not raised until a century ago in his well-known book, *The Coal Question*. In it W. S. Jevons, a prominent nineteenth-century English economist, calculated the duration of coal supplies used to power England's industrial might. As it happens, Jevons was mistaken in projecting that without greatly increasing the price of fuel England would soon exhaust its reserves of coal. Like Jevons, later scholars who attempted to estimate the magnitude of energy resources were proved wrong as well, as vast new deposits continued to be discovered.

The discovery of oil is conventionally dated to 1859, when the first successful well was drilled in Pennsylvania in the deliberate search for an oil reservoir (see Plate 1). Obviously, at the time there was no concern about oil's long-run supply. And when supply questions were raised, the task of estimating how much proved an exceedingly difficult one.

There are several reasons for this. Most early coal discoveries showed at the surface or were situated at a shallow depth; thus they were easily measured, at least approximately. Oil and natural gas, except for those rare occasions when there is surface seepage, are wholly hidden from view and often located far below the surface. Moreover, they are not found in pools but exist as drops of liquid or bubbles of gas in the pores, or interstices, of various types of rock. They emerge first under internal, natural pressure, when the drill opens a passage to the surface. When that pressure is exhausted, they must be coaxed out by pumping or by other artifical means.

Thus, many factors need to be assessed in order to estimate the size of an oil or gas discovery. The most important are the geologic indicators—the dimensions and porosity of the rock formation which contains the oil or gas, and the degree to which the characteristics of the rock will permit escape of the liquid or gas—technically known as its permeability. Moreover, engineering judgment and luck play an important role. Except when developing a known field, it is easier for a drill to miss a reservoir than to find it; statistics show failure is far more frequent than success. So, at times, old drill sites, abandoned as dry holes, have been revisited, and oil or gas has been located where a prior attempt has failed. Until recently, for example, the Overthrust

Belt, now one of the most promising oil and gas provinces in the country, was called the "well-driller's graveyard."

In short, the search for oil and gas is a very chancy business, and such uncertainty plays havoc with estimates of resources and supply predictions. At different times experts have either foreseen the early exhaustion of oil and gas resources—and have been proved wrong—or predicted exceedingly large potential resources—and have been proved equally mistaken (though there can be no convincing direct proof in the latter case; it must be by indirection).

TODAY'S CONVENTIONAL ENERGY RESOURCES

In 1920 the chief geologist of the U.S. Geological Survey (USGS) estimated that there were only about 7 billion barrels of oil to be recovered in the United States by then-available technology. It was highly improbable, he added, that the error in that judgment exceeded 50 percent. At the rate of consumption— then some 500,000 barrels per day (or about 3 percent of what it is now)— U.S. petroleum sources probably would be exhausted in fourteen years. Sixty-one years later 130 billion barrels have, in fact, been recovered, crude reserves amount to nearly 30 billion barrels, and the USGS believes that there is a fifty-fifty chance that another 83 billion barrels could be discovered and recovered. Illustrating the wide latitude are three studies that, in 1962, attempted to assess undiscovered recoverable petroleum resources in the United States: the lowest estimate put the quantity at 91 billion barrels, the middle one at a range of 289 to 399 billion, and the highest one at 648 billion barrels (Table 3-1).

Notwithstanding such disconcertingly different estimates, the urge to know persists and one does the best one can with the knowledge and methods at hand. It is therefore appropriate to briefly discuss the concepts that underlie resource estimates together with displaying figures that follow from them.

The discipline that concerns itself with estimating energy resources makes a distinction between reserves and resources. *Reserves* are defined as material (coal, oil, gas, and so on) that is known to exist in a given location, to be of a calculable magnitude with only a moderate degree of uncertainty (say, 20 percent plus or minus), and to be profitably and, of course, in the individual case, legally recoverable through the use of existing technology. One may further distinguish between *proved, indicated,* and *inferred* reserves. As the terms suggest, the sequence is one of decreasing certainty.

In this hierarchy, even inferred reserves are far more "real" than the next category, *potential resources,* which are conventionally called "undiscovered recoverable resources" (the category dealt with in Table 3-1). Estimates here are not typically the result of drilling, but of other types of information, such as geologic knowledge and inferences from past discoveries. The term *resources* is usually applied to the entire endowment, including reserves.

The Dilemma of Resource Use

There is a widespread notion that natural resources are a treasure to be left unexploited until "needed," and that extracting and consuming them is an act of thoughtlessness and waste. Leaving aside the fact that resources can indeed be "wasted," that is, consumed at a price that does not reflect their full cost, the notion of equating use with waste derives from a failure to grasp that resources constitute an asset. Like any other asset, resources are converted, by use, into services and into other assets. Fuel burned in the manufacture of, say, steel, helps to shape assets like factories or tools. Fuel burned in farm machinery or turned into fertilizer to grow crops keeps people alive and strong enough to become skilled workers. More assets are created. Indeed, one of the perplexing issues confronting a rich, oil-exporting country like Saudi Arabia is how to convert an oil-based, single-product economy, into a modern, diversified, industrialized country. Sitting on the oil will not do it. Selling it and using the proceeds to build a modern economy will.

TABLE 3-1.   Variations in Estimates of U.S. Petroleum Resources,
for Selected Years, 1959-81 (in billions of barrels)

Date of estimate	Source	Undiscovered recoverable resources
1959	USGS Bulletin 1136	387
1960	Weeks	74
1962	National Academy of Sciences	91
1962	National Fuels and Energy Study	289-399
1962	Energy Policy Staff—DOI	648
1965	USGS Circular 522	264
1972	USGS Circular 650	450
1974	Mobil	88
1974	Hubbert	67
1975	USGS—Resource Appraisal Group	98
1975	National Academy of Sciences	113
1975	USGS Circular 725	50-127
1981	USGS	64-105
1981	Nehring	14- 32

Note: To be comparable, the estimates, made at different times, are adjusted to 1974 as a common basis. Details of individual sources may be found in publications cited in source note.

Source: Hans H. Landsberg, and coauthors, *Energy: The Next Twenty Years,* A Study Sponsored by the Ford Foundation and Administered by Resources for the Future (Cambridge, Mass., Ballinger, 1979); and U.S. Geological Survey, "Estimates of Undiscovered Recoverable Resources of Conventionally Producible Oil and Gas in the U.S., A Summary" (Reston, Va., 1981) p. 6; and Richard Nehring, *The Discovery of Significant Oil and Gas Fields in the United States,* Prepared for the U.S. Geological Survey and the U.S. Department of Energy (Santa Monica, Calif., Rand Corporation, 1981) p. 175.

Some years ago, Vincent McKelvey, then-director of the USGS, prepared a diagram showing these classifications. Resource students now know this as the McKelvey box. Although subsequently modified, its original form is a straightforward way of conveying the essential features of resources and reserves. Two slightly expanded forms, both prepared by the U.S. Bureau of Mines, are shown in Figure 3-1. The reserve portion, from which current production is drawn and which is thereby continuously depleted, receives reinforcements from *below,* as economic conditions (for example, higher prices) or improved technology (for instance, deeper-reaching drilling equipment) turn marginal resources into reserves, and from *the right,* as the degree of certainty increases and new deposits are discovered. That is to say, while the diagram depicts a one-time situation, it is intended to suggest that there is constant flux between the compartments.

One thing to bear in mind when confronted with data on reserves and resources is that no matter how precise they may look, they are merely estimates—that is to say, they are the product of someone's judgment. This judgment will vary among estimators and over time. Yet, because the need to know is pressing, figures will be produced, and we should resign ourselves to the

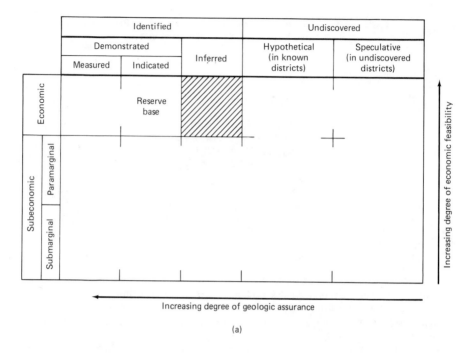

Increasing degree of geologic assurance

(a)

Cumulative production	Identified resources			Undiscovered resources		
	Demonstrated		Inferred	Probability range		
	Measured	Indicated		Hypothetical	(or)	Speculative
Economic	Reserves		Inferred reserves			
Marginally economic	Marginal reserves		Inferred marginal reserves			
Subeconomic	Demonstrated subeconomic resources		Inferred subeconomic resources			

Other occurrences	Includes nonconventional and low-grade materials

(b)

Figure 3-1 The classification of resources. (a) An early version. *Source*: U.S. Bureau of Mines, *Mineral Facts and Problems*, Bicentennial Edition (Washington, D.C., 1976). (b) From *Energy: The Next Twenty Years*, © 1979, The Ford Foundation. Reprinted with permission from Ballinger Publishing Company.

fact that users of the estimates will continue to consider them more meaningful than they are. With this cautionary preface, how do we assess the dimensions of energy reserves and resources today? Starting from the most visible source, what is the outlook for coal?

Coal

United States. A concept has long been utilized in the United States that relates to the standard terminology discussed above but does not quite conform to it: the *demonstrated coal reserve base.* This denotes the amount of coal in place that corresponds to specified criteria of depth and thickness of seam (not more than 1,000-feet deep and at least 28-inches thick for good-quality coal), for which there exists a high degree of geologic information and engineering evaluation, and that can be recovered economically and legally, with actual recovery depending on the individual circumstances. In other words, this is a definition of the amount in place. How much will, in fact, be recovered depends on mining methods and other factors and may range from 40 to 90 percent of the estimated amount. The higher rates occur in surface mining and the lower in underground mining, where most methods lead to substantial amounts of coal being left behind. As shown in Table 3-2, 33 percent of all U.S. coal is suitable for surface mining; most of it is located west of the Mississippi River, in the more recently worked coal areas of the Northern Great Plains, especially Wyoming, Montana, and North Dakota. Still, the Western Region contains slightly more underground than surface coal. In contrast, the East overwhelmingly contains coal accessible only by underground mining methods.

The usefulness of coal is largely determined by its sulfur content and its heating value. The higher the sulfur content, the more troublesome the use, as special equipment is needed to reduce the amount of sulfur that, upon com-

TABLE 3-2. U.S. Coal Reserves, 1979 (billion short tons)

Type	East	West	Total
Surface	41.8	114.6	156.4
Underground	172.5	145.7	318.2
Total	214.3	260.3	474.6
Percentage with low sulfur	20	70	51

Sources: U.S. Department of Energy, *Demonstrated Reserve Base of Coal in the United States on January 1, 1979* (Washington, D.C., May 1981). Sulfur content percentages are from Sam H. Schurr, Joel Darmstadter, Harry Perry, William Ramsay, and Milton Russell, *Energy in America's Future: The Choices Before Us* (Baltimore, Md., Johns Hopkins University Press for Resources for the Future, 1979) p. 228.

bustion, escapes up the stack. Here the western states have a great advantage, though, as noted below, not as great as sometimes has been thought. About 80 percent of western coal contains less than 1 percent sulfur, a level that is considered next to benign. In contrast, less than 30 percent of eastern coal has that low a sulfur content. At the opposite end of the range, 20 percent of eastern coal contains more than 3 percent sulfur, while only a little over 1 percent of western coal falls in that range.

Differences in heating value, that is, the amount of heat that is generated in combustion, are also substantial. Eastern coal generally has a high heating value, ranging from 10,500 to 14,000 Btu per pound; western coal, which is high in ash and moisture content, ranges from only 6,300 to 11,500 Btu. A special place is occupied by some West Virginia coal which is among the few low-sulfur coals found in the East and which additionally lends itself well to producing coke—a strong, chunky, porous material, from which virtually all components but carbon have been eliminated through heating. Such coal is required in steel making to provide the heat, gas, and structural support needed to produce iron from iron ore.

Heating value and sulfur content are, of course, related in that the advantage of the low-sulfur value of western coal is significantly offset by the fact that its low heating value compels the user to burn more of it to obtain a given quantity of energy. Thus more sulfur is emitted per unit of heat (as compared to unit of weight). If a low-value coal, say, of 7,000 Btu per pound with 1-percent sulfur content, is burned, it would produce the same amount of heat and sulfur as a half pound of coal with 14,000 Btu per pound with 2 percent sulfur.

As shown in Table 3-2, the U.S. demonstrated coal base measures roughly 475 billion tons. Assuming conservatively that only half of it can be recovered, the resulting 237.5 billion tons are equivalent to about 350 times current annual U.S. consumption of coal. Even on the assumption of increased use of coal, that is a comforting margin—all the more so when one considers that the demonstrated reserve base is only a small fraction of total resources.

It is entirely likely that these figures exaggerate reality. Not every ton considered recoverable will be produced. Special locational problems, environmental difficulties, political impediments, ownership patterns, and so on will whittle away at these figures. Moreover, experts have made widely different estimates, all based on reasonable assumptions, none of which are verifiable. Nonetheless, few would argue that we need to worry about a shortage of U.S. coal. If coal use were to be constrained, it would be for reasons other than resource limitations.

More intriguing is the question of where coal will be mined—in the East or the West? (So-called Interior coal, that is, coal from the midwestern tier of states, from Indiana and Illinois in the North to Texas and Oklahoma in the South, occupies a somewhat intermediate position, but in most respects it is more akin to the East than the West, especially with regard to its sulfur content.)

Western coal has overcome the handicaps of its low heating value and high shipping costs and competes successfully in midwestern and south central markets because it is cheap to mine (by surface) and cheap to burn (because of its low-sulfur content). It remains cheaper to burn even though air pollution regulations now demand that all combustion gases, regardless of the original sulfur content of coal, be subjected to a cleaning process (by the use of "scrubbers" to remove much of the sulfur in the stack gas). While it is impossible to prove the point, the data for coal production by region suggest that, if these regulations were designed to work to the disadvantage of western coal, that resource nevertheless has had a lusty growth. Because most western coal is mined by labor not organized by the United Mine Workers it has by and large been free of the periodic strikes common to eastern coal fields and this factor may have enhanced further its value to customers.

Worldwide. Coal is abundant not only in this country, but over much of the rest of the world. Statistics on coal resources, however, need to be evaluated carefully. For example, in the most authoritative compilation of such data (see Table 3–3), U.S. coal reserves are listed at only 113 billion tons, that is, one-fourth of the reserve base estimate, and one-half of the recoverable

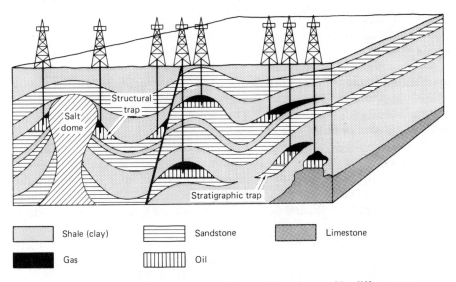

Figure 3-2 Types of oil and gas accumulations. Although caused by different natural phenomena, all types of hydrocarbon traps perform the same role. They provide a site for the subsurface accumulation of oil and gas and thereby create the fields that are sought by the drill. Note examples of major traps. This material is reproduced with permission from *How Much Oil and Gas?*, Exxon Background Series, © 1982 Exxon Corporation.

TABLE 3-3.   World Resources of Bituminous Coal and Anthracite
(million tons of coal equivalent)

Continent and country	Geological resources	Reserves
Africa	172,714	34,033
Botswana	(100,000)	(3,500)
South Africa	(57,366)	(26,903)
North America	1,308,541	126,839
United States	(1,190,000)	(113,230)
Canada	(96,225)	(8,706)
Asia	5,494,025	219,226
People's Republic of China	(1,424,680)	(98,883)
USSR	(3,993,000)	(82,900)
Australia and South Seas	213,890	18,164
Australia	(213,760)	(18,128)
Europe	525,664	94,210
Federal Republic of Germany	(230,300)	(23,919)
Poland	(121,000)	(20,000)
United Kingdom	(163,576)	(45,000)
Total	7,724,834	492,472

Note: "Equivalent coal" has a heating value of 25.2 million Btu per short ton. It is a convenient, typical value for coal to which all coals of different heating value are adjusted. Numbers shown in parentheses reflect only the resources of the country with the largest reserves within that continent and are not addable to the totals.
Source: World Energy Conference, World Energy Resources, 1985–2020 (New York, IPC Science and Technology Press, 1978) pp. 66–67.

quantity, if we assume a 50 percent recovery rate. The reason for this is that (1) only recoverable amounts are shown (not coal in place); and (2) low grades, such as subbituminous coal and lignite, are excluded. This seems a poor approach, since it excludes, for example, all western U.S. coal which has been providing a rapidly rising share of U.S. coal. But since the absolute amounts are very large, such factors do not substantially vitiate the general picture of adequacy. World reserves—that is, recoverable coal resources commercially viable now—are estimated at some 500 billion tons. Compared with an annual consumption rate of about 3 billion tons in 1980, this is a handsome endowment.

Coal is widely distributed, and it probably exists in a good many more places not yet identified because the need to explore more widely has not arisen. The Soviet Union contains about half the world's coal resources and not quite 20 percent of the reserves. China is next, with about 20 percent of both resources and reserves; and the United States is third, with about 15 percent of resources and 23 percent of the world's reserves (not counting subbituminous coal and lignite). About twelve countries produce more than 90 percent of the world's coal, and the top three producers—the United States, the Soviet Union, and China—account for close to 60 percent. This is an interesting, and perhaps surprising fact, since one thinks of concentrated production as being more characteristic of petroleum than coal. Yet, the three nations leading in the production of crude oil accounted for 50 percent of world output in 1980, and the top twelve for 80 percent. In both cases, concentration is lower than in the case of coal. However, the big difference in impact is that coal enters much less into international trade, and thus concentration in coal does not affect reliability of supply as is the case in heavily traded oil.

Petroleum

United States. The uncertainties that afflict estimates of coal reserves and resources are multiplied when it comes to oil. While coal exists as solid layers to be broken up or scooped out, oil exists in tiny droplets embedded in rock or highly compacted sand (see box on page 62). As explained earlier, an oil field is not a pool of liquid, but a rock layer that in various ways contains pockets of oil, as is illustrated in Figure 3-2. This has a number of consequences. Oil is harder to discover; being liquid it is fugitive, that is, it is not fixed in place; it may exist at several depths in one and the same location; it may give few clues as to its horizontal dimensions; it varies greatly in ease of recovery; and, since it is costly to explore, oil companies have refrained from investing more in exploration than is required to provide an adequate production horizon—usually not more than ten years ahead.

Estimates of reserves have often mistakenly been taken to mean "all there is to work on." Thus people have been puzzled to discover that as production has continuously increased, so have reserves, at least over most of the history of oil and gas. Especially for the world as a whole, additions to reserves have been spectacular. It was only since the 1970s that additions to U.S. reserves from new discoveries or extensions of old ones have failed to keep pace with production. Total reserves began declining in 1971. In terms of the McKelvey box (see Figure 3-1), it means that there has been insufficient movement from the *lower right* to the *upper left*. The largest single injection of new reserves into the U.S. picture was the Alaskan North Slope find of 1968, which added about 9 billion barrels. Nothing comparable has occurred since, but motivated by steeply rising crude oil prices, drilling for both new fields and extensions of known fields rose sharply, first slowing down the decline in U.S. crude reserves and reversing it in 1981. As might have been expected, the drilling boom subsided quickly as oil prices stopped rising in mid-1981. Table 3-4 shows the pertinent data.

Users of resource statistics need to know that data may differ with respect to the inclusion or exclusion of natural gas liquids (NGL), a material stripped from natural gas that refiners use jointly with conventional crude. These so-called NGL reserves total about 6.7 billion barrels. Thus, oil reserves,

The Requisites for Recoverable Petroleum to Exist in a Sedimentary Basin*

For oil and gas to have accumulated in a reservoir and to be recoverable, all of the following had to be present:

- A *source rock,* such as a shale, which was formed millions of years ago when layers of mud containing billions of microscopic animals and plants sank to the bottom of a sea or lake. Over time, as additional layers of sediments built up on top of them, these muds were subjected to heat and pressure, causing them to solidify and form shales. This process also led to the decomposition of the organisms deposited with the mud into compounds of hydrogen and carbon that we now know as petroleum.
- A *reservoir rock,* such as a sandstone or limestone layer, which was in contact with source rocks so it could receive petroleum that was squeezed out of the source rocks as they were compressed by overlying sediments. The reservoir rock had to be porous—have cracks and open spaces between its grains—to provide a place to store the petroleum received from the source rocks. It also had to be permeable so that the oil and gas could move through it.
- A *trap* (see Figure 3-2), where oil and gas—which are lighter than the water that is also in the reservoir—have accumulated. There are two general types of traps: *structural* traps, formed by the folding or faulting of the reservoir rock layers, and *stratigraphic* traps, created when layers of porous, permeable reservoir rocks are sealed off by superimposed impermeable beds.
- A *seal,* which is the dense rock that overlies the trap and prevents the petroleum liquids and gases from escaping.

Not only must all these features have been present; they also must have been properly related in time. The source rock had to be in contact with the reservoir rock after the organic matter was transformed into hydrocarbons so that they could be squeezed out of the shales and into the reservoirs. Similarly, the traps (and the seals) had to exist at the time the hydrocarbons were being forced into the reservoir rocks, in order to limit the movement of the oil and gas and so cause them to accumulate in what we today call oil and gas fields.

*This material is reproduced with permission from *How Much Oil and Gas?,* Exxon Background Series, © 1982 Exxon Corporation (New York, Exxon Corporation, May 1982).

TABLE 3-4. U.S. Proved Reserves of Crude Oil and Natural Gas Liquids, for Selected Years, 1950–81 (in billions of barrels)

	Reserves	
Year	Crude only	Crude plus natural gas liquids
1950	25.3	29.6
1955	30.0	35.4
1960	31.6	38.4
1965	31.4	39.4
1970	39.0	46.7
1975	32.7	39.0
1976	30.9	37.3
1977	31.8	37.8
1978	31.4	38.2
1979	29.8	36.4
1980	29.8	36.5
1981	29.4	36.5

Source: American Petroleum Institute (API) data, as reported in *1980 Annual Review of Energy,* p. 41, for years prior to 1977. Data for subsequent years are from U.S. Department of Energy, *U.S. Crude Oil, Natural Gas, and Natural Gas Liquids Reserves, 1980 Annual Report* (October 1981) p. 10. The 1977 figure for natural gas liquids is also from the API source.

as of the end of 1981, are variously listed as 29.4 billion barrels or as 36.5 billion barrels, the first number excluding, and the second including, natural gas liquids. The latter is the more common practice, since the two substances are so closely related.

Estimating the amount of undiscovered oil that remains to be recovered has long intrigued geologists, economists, and, of course, policymakers. It was once assumed that experience in known oil-producing areas could be widely replicated. This assumption naturally yielded very large resource estimates. More recently, estimates have been based on specific reservoir-by-reservoir approaches, in which detailed geologic analogy plays a large part. Correlation between the presence of certain classes of rock, certain configurations, and the presence of oil has formed the basis of conjecture as to locations of hitherto undiscovered oil; confirmation of the hypotheses, however, can be proved only by drilling. Probably the most costly failures have occurred offshore, where companies at times have invested hundreds of millions of dollars in exploratory drilling only to come up with dry holes. Still, new areas continue to turn up, the latest being the so-called Overthrust Belt, a long north-south zone stretching from Alaska to Mexico that has been the site of significant discoveries. Intensive and successful exploration in the early 1980s for both oil and natural gas has concentrated in Wyoming and Utah.

Energy to Produce Energy

There are many ways of producing energy, all of which involve application of energy or heat; some more than others. Two examples will suffice. One is enhanced oil recovery (see page 65). In this instance, chemicals or heat must be injected into an oil-bearing formation that, beyond a certain point, refuses to yield up the liquid held in its pores. The cost of this heat or energy has gone up with the cost of oil. Thus recovery processes have become more expensive, and the balance of advantage has become dubious. Another example is heavy crude oil, such as that which exists in vast amounts in parts of Venezuela and, to a much smaller extent, in California. To coax any of this oil to the surface requires the heavy application of heat. Again, it is a question of balance between the cost of energy to be injected, plus other operating costs, and the price to be obtained for the oil as it gains in fluidity and rises to the surface or as it is pumped out.

In a more general sense, questions have been raised as to the extent to which a given source of energy contributes "net" energy, that is, whether the energy output exceeds the energy input, when the input is calculated to include both direct energy and indirect energy incorporated in capital equipment, transportation, and so forth, used in production. While it is theoretically possible to expend more energy than is eventually produced, in a market economy this would be commercially successful only if the energy produced had a higher value (gasoline, for example) than the energy employed in production. Basically, what we have here is the contrast between a physical and an economic calculus. In the real world, the latter will shape decisions, but, at times and especially in close decisions, it can be instructive to look at both.

While reserves are conventionally presented as single numbers, such as the above-mentioned 36.5 billion barrels, in recent years resources increasingly have come to be described in probability terms. The most detailed such estimate was made first in 1975 and revised in 1981 by the USGS. It suggests that, on the basis of a careful field-by-field survey, there is a 95 percent chance that undiscovered, recoverable resources would amount to at least 64 billion barrels and only a 5 percent chance that they might be larger than 105 billion barrels. An often-cited middle estimate is 83 billion barrels. A breakdown of these estimates is shown in Table 3-1. As U.S. petroleum consumption ran at an annual rate of not quite 6 billion barrels in the early 1980s, the reserve estimate for domestic crude equaled a six-year supply at that consumption level if wholly met by domestic production; the middle resource estimate implied an additional fourteen-year supply.

Neither of the two calculations is very reassuring. To be sure, domestic oil will be supplemented by imports and stretched by future fuel-switching and conservation, so that domestic crude supplies will last beyond the end of the century. But the beginning of the twenty-first century is not that far away, and the outlook is unlikely to improve as long as the United States continues to consume oil at such a high rate. At the consumption level of 6 billion barrels per year, it would take the discovery of several giant fields to increase U.S. reserves significantly. Yet, there are no indications that such fields are likely to be found, either onshore or offshore. Note that when the giant Prudhoe Bay field on Alaska's North Slope was discovered, it was hailed as a turning point in U.S. petroleum history; and, indeed, in light of the 4.9 billion-barrel annual domestic consumption at the time, it was a very large addition. Although it would still be considered large today, it would not be looked upon as reversing the gradual erosion of domestic oil production, when annual consumption has moved to 6 billion barrels in the short span of only a decade.

There is one source of crude oil that is much larger than those undiscovered amounts we have talked about, and that is the oil left behind in fields that already have been tapped. In the past, generally one-third of the oil in place in all reservoirs tapped has been recovered. It follows that twice as much oil has been left behind. With roughly 130 billion barrels produced between 1859 and 1981, there must then be some 260 billion barrels left in reservoirs whose locations are known. In a sense, this would be by far the largest "reserve" in the United States. However, attempts to extract this oil have not been very successful. There are no statistics at this point, but estimates of what might reasonably be achieved do not anticipate that more than 10 to 30 billion barrels altogether can be recovered, that is, 10 percent or less of all the oil left behind. And these are goals, not experienced facts.

Enhanced Oil Recovery (EOR) has been the object of industrial and, lately, also governmental effort and expenditures. But it has remained an elusive goal. Industry sources blame differences in physical and chemical characteristics of reservoirs for keeping a technique that is successful in one place from working elsewhere. Experimentation is costly, and with the rising price of oil, the petroleum-based materials that are often needed for EOR have also become more costly. Nonetheless, the sheer volume of oil known to exist in no-longer productive reservoirs is so large that efforts to recover it are bound to continue to intrigue inventive epxerimenters, and perhaps some day the key to success will be found. For the time being, however, it remains more a fond hope than a significant contributor to production.

Worldwide.   While the outlook for U.S. reserves and resources seems rather bleak, the worldwide picture is somewhat more encouraging. Global reserves are estimated at about 650 billion barrels, and undiscovered recoverable resources about twice that size. In the late seventies worldwide consumption of oil amounted to about 22 billion barrels a year. By the early 1980s, that figure had declined to a little more than 20 billion barrels. Thus world reserves are equivalent to some thirty years' current consumption—more, if consumption continues to decline as it has done recently, but less if it picks up. The additional 1,300 billion barrels or so of potential resources present a more comfortable horizon—some sixty to seventy additional years at recent consumption levels; but that figure is highly speculative. Over the past four decades, estimates gradually rose from some 400 to range between 1,800 to 2,220 billion barrels, before stabilizing. Even so, one does well to regard them as grossly approximate figures.

The numbers do suggest that the "oil age" is by no means over. Those holding a more sanguine view point to the discoveries in the North Sea and in Mexico. Between them they now produce some 5 million barrels a day, where ten years ago they produced practically nothing, and neither area is likely to have reached its final magnitude. They also note that there has been very little, if any, drilling in many parts of the world, and that it is, therefore, too soon to close the book on petroleum discoveries. Those holding more pessimistic views counter that where there has been little drilling, it is because geological information suggests lack of success; as for North Sea and Mexican oil resources, these are indeed large new discoveries, but between them they contribute less than 10 percent to current world supplies, and thus do not much alter the essentially downward trend in replenishment of oil resources. Finally, United States production is not likely to rise, with new discoveries at best offsetting continued depletion of oil fields.

Natural Gas

United States.   What has been said for petroleum trends holds in many ways for natural gas, but there are differences. Sometimes found in association with petroleum and sometimes by itself, natural gas has been looked for only in the relatively recent past, and, on an international scale, only in relatively few countries, mostly in the United States. A major reason why gas is a latecomer is that it is best transported by pipelines. (Transportation in compressed liquid form, a relatively recent development, is expensive, and can be used commercially only with very large quantities of gas.) Even in the United States, serviceable pipelines were not built until the late twenties, and improved only in the 1940s and 1950s to make long-distance movement feasible. For a long time, much of the gas recovered jointly with petroleum found no local use, had no pipeline connections, and was disposed of by burning.

As is true for petroleum, U.S. reserves of natural gas are declining, having reached their peak in 1967. Since then, except for 1970, extraction has outpaced additions to reserves. By 1980, reserves were equivalent to less than ten years of consumption at current levels. Undiscovered recoverable resources were estimated at the equivalent of a twenty- to thirty-five-year supply. These time horizons are not significantly longer than those for domestic crude oil, especially if the margins of error are taken into account. Table 3–5 shows how steeply these estimates have declined during the past two decades, and especially in the last dozen years or so.

Recently, there has been much talk of large future increases in natural gas resources. It was touched off by sizable discoveries at depths of 15,000 feet or more, deeper than levels at which gas had previously been looked for.

TABLE 3-5. Estimates of U.S. Natural Gas Resources,
for Selected Years, 1959–81 (trillions of cubic feet)

Date	Source	Undiscovered recoverable resources of natural gas
1959	USGS Bulletin 1136	1,804
1962	National Fuels and Energy Study	1,250
1962	Energy Resources Report	800
1962	Department of Interior to AEC	2,318
1965	USGS, Circular 552	1,080
1972	USGS, Circular 650	2,100
1973	Potential Gas Committee	880
1974	USGS	1,000–2,000
1974	Mobil	443
1974	Hubbert	361
1975	NAS	530
1975	USGS, Circular 725	332–655
1981	USGS	474–739
1981	Nehring	143–209

Source: Hans H. Landsberg, and coauthors, *Energy: The Next Twenty Years,* A Study Sponsored by the Ford Foundation and Administered by Resources for the Future (Cambridge, Mass., Ballinger, 1979); Richard Nehring, *The Discovery of Significant Oil and Gas Fields in the United States,* Prepared for the U.S. Geological Survey and the U.S. Department of Energy (Santa Monica, Calif., Rand Corporation, 1981) p. 176; and U.S. Geological Survey, "Estimates of Undiscovered Recoverable Resources of conventionally Producible Oil and Gas in the U.S., 1981, A Summary," (Reston, Va., 1981) p. 6.

Second, drilling in the Overthrust Belt, explored in the past without significant results, suggests not only that a large quantity of gas might be obtained in that region, but also that the country has not come near the end of discoveries. It is too soon to consider that these expectations have been tested convincingly but, in view of the shorter history of natural gas exploration in this country, it is a plausible hypothesis. As in the case of oil, however, current levels of consumption are so high that new discoveries would have to be extraordinarily large to change the outlook for domestic gas resources. That is why the so-called unconventional gas resources have aroused so much interest, as noted below.

Worldwide. Global figures are not very instructive, both because gas consumption is limited in the main to the industrialized countries, and the search for gas has not been extensive as yet. Starting in 1967, worldwide data on proved gas reserves were published, and within fifteen years these estimates had doubled. One reason for this is that in most parts of the world natural gas was considered a nuisance to be disposed of. Large Middle Eastern oil producers only recently have begun to pump back into the ground for later recovery much of the gas they extract and are in the process of establishing chemical facilities that will use it. Still, much of the Middle East continues to flare its natural gas.

World proved reserves would meet the equivalent of forty to fifty years' equivalent of current consumption. Estimates for ultimately recoverable gas reserves worldwide vary but in recent years have ranged from 6,000 to 10,000 trillion cubic feet. Even the lower range is equivalent to at least one hundred years of current world consumption. Regionally, these resources are distributed in a manner similar to petroleum; that is, the United States, the Middle East, and the Soviet Union contain the principal natural gas deposits. The Soviet Union has become a growing exporter of natural gas, especially to Western Europe, where it supplements local supplies from Netherlands fields, the North Sea, and small local sources elsewhere.

Because 80 percent or more of the natural gas in gas-bearing formations is extracted when gas reservoirs are tapped, the matter of enhancing recovery, so important in oil, does not arise in natural gas. Thus future reliance must be solely on those deposits currently undiscovered and those sources not yet commercially used or significant.

Uranium

Uranium is the latest entry in the catalog of energy sources. Its existence has been known for a long time, but its use has been limited to a few very specific applications, such as luminous watch or instrument dials, X rays, and a few others. None were on a large enough scale to call for an investigation into adequacy of this resource. Incidentally, just as little was known about the insidiousness of its radiation characteristics.

What changed all this was the birth of the nuclear age that can be dated from the first controlled chain fission reaction, which took place in December 1942 on a squash court at the University of Chicago. This opened up the possibility that the world might solve its energy problem through the use of uranium. One ton of this material has the energy potential of about 3,500 tons of coal in the current type of reactor, and at least 60 times that much when used in a breeder reactor. It was clear at the same time that the path to the breeder was long and that the initial class of nuclear reactors would be very wasteful of uranium. Added to the heavy demand of the military for bomb material, the prospective large-scale use of uranium for electric power generation created the need to inquire into the extent of uranium reserves and resources.

A discussion of uranium reserves and resources poses two sorts of inconvenience. One is methodological and consists in the fact that, for the United States, a terminology has developed that introduces yet one more set of reserve and resource concepts and classifications. Moreover, they also differ from what the rest of the world uses. The second and far more profound drawback is that the significance of the numbers depends enormously on the technology in which the uranium is used. Any quantity that would last, say, fifty years, when used in the current fission reactors—those using only 0.7 percent of each ton of uranium—would last some 3,000 years when used in a breeder reactor. The reason for this is that the breeder reactor can convert most of its uranium feedstock into fissionable material (see Chapter 4).

Uranium resource data have one further peculiarity: the estimates are linked to assumptions about costs of uranium production. It is a useful concept that one would like to see utilized more widely, but it creates problems because the meaning of specific prices changes over time and because, in this instance, costs are defined somewhat differently in the United States than in the rest of the world.

Table 3-6 displays the latest data. Cast in terms of uranium oxide (U_3O_8)—also called yellowcake, the chemical form in which uranium energy emerges from milling—they show that the world's "reasonably assured" resources (roughly similar in meaning to reserves) range between 2.3 and 3.0 million tons, depending on the cost assumed per pound. This approximates between 100 and 140 years at current levels of uranium consumption. Estimated "additional resources" which correspond roughly to what U.S. uranium statistics call "probable" and "potential" resources, based on different criteria of geologic evidence, total about once again as much as the reasonably assured resources.

The worldwide distribution is very different from that of coal, oil, and natural gas. The United States is far in the lead, a reflection of the uranium

TABLE 3-6.   World Uranium Resources (1,000 short tons of U_3O_8)

Country	Reasonably assured		Estimated additional	
	At production cost of $30 or less	At production cost of $50 or less	At production cost of $30 or less	At production cost of $50 or less
South Africa	321	463	109	228
Australia	382	412	343	371
Canada	299	335	465	988
Niger	208	208	69	69
Namibia	155	176	39	69
Brazil	155	155	106	106
Other, excluding the U.S.[a]	280	445	70	279
United States	471	787	885	1,426
TOTAL	2,271	2,981	2,086	3,536

Note: Production cost for this purpose includes only costs subsequent to land acquisition and exploration.

Source: Energy Statistics, vol. 5, no. 2 (Second Quarter, 1982), a journal of the Institute of Gas Technology, Chicago, Ill.

[a] Excludes USSR, the People's Republic of China, and Eastern Europe.

boom of the fifties and sixties, when government incentives established to assure sufficient supplies for military purposes sent prospectors armed with Geiger counters roaming the West in the search of the new material. They found it, mostly in the Rocky Mountain states, almost exclusively in shallow sandstone formations. The Republic of South Africa, Australia, Namibia, Canada, Brazil, and Niger hold other large uranium reserves. In potential resources, the United States and Canada dominate the picture.

The adequacy of uranium resources was a hotly debated issue in the 1970s, with the views of many participants in that debate significantly shaped by their judgment of whether or when the nuclear breeder should be introduced into the energy economy (see Chapter 4). Those advocating need and speed tended to foresee inadequate and expensive uranium supplies here and abroad. Those taking a more relaxed or outright hostile attitude toward the breeder tended to stress factors suggesting that adequate supplies of reasonably priced uranium would come forth as needed. With this background in mind, one may say generally that some believed that:

• The resource estimates were highly conservative and, like all other resource estimates, they were bound to be boosted in time.

• Higher prices would call forth larger reserves, both in conventional geologic environments and in new ones, without significantly affecting the cost of nuclear-generated electricity.

- In any event, the rate of growth of the nuclear power industry was going to slow down.

Others believed, just as firmly, that:

- The estimates had a large margin of uncertainty.
- The chances for overestimating were as great as those that understated reality.
- It was prudent to assume the former.
- The price of uranium was destined to rise (beyond the level of $40 a pound at which U_3O_8 was then being traded).
- The existence of reserves did not guarantee timely supply in needed volume.
- In short, the industry was in trouble, unless steps were taken to move rapidly to recycle used-up reactor fuel as well as to proceed with development of the breeder.

In the space of only a few years the debate has become moot, not because uranium data are better or that vast new quantities have been found (though there have been substantial discoveries), but because the slowdown in the growth of nuclear power leaves little doubt that adequacy of uranium is not a pressing problem. Uranium in the early 1980s was in long supply, its price had fallen significantly, so that in constant (1981) dollars it was worth about half of what it was worth five years earlier. A strong revival of nuclear power could rekindle the debate, but it is not now in sight, even though some countries—especially France and the Soviet Union—have not encountered the letdown that has depressed the U.S. industry.

UNCONVENTIONAL OIL RESOURCES

Since liquid fuel has caused consumers more trouble than other forms of energy, the search for direct petroleum substitutes in recent years has received the greatest emphasis. Alternative resources are oil shale and tar sands; and heavy oil, a hard-to-produce form of crude oil, occupies an intermediate position. Another substitute, liquefied coal, is discussed only in Chapter 4, since it is not an unconventional *resource* but rather the product of an unconventional *conversion process*.

Shale Oil

Shale oil is attractive because its parent material, oil shale, is abundant. Even counting only high-grade oil shale (that is, shale with a high oil content),

How Long Will U.S. Uranium Last?

A good rule of thumb is that a 1,000-megawatt (MW) reactor of the kind now in use in the United States uses about 200 tons of uranium oxide (U_3O_8) per year, assuming average operating experience and no recycling of spent fuel. If the reactor lasts thirty years, it will require 6,000 tons of U_3O_8 to keep it producing electricity. Technological improvements and recycling could substantially lower that figure. But sticking with it for simplicity's sake, the 55,000 MW of electrical capacity of U.S. nuclear power plants in 1982 require 11,000 tons of U_3O_8 per year, or roughly 330,000 tons over their lifespan. That equals less than half the 790,000 tons of U.S. uranium reserves, up to $50 per pound. Put differently, reserves either could support for thirty years (to 2013) a nuclear power industry more than twice as large as that which existed in 1983 (or an industry of the present size for more than seventy years). Such a capacity is likely to be reached or exceeded by the end of the century. Since capacity growth between now and the future will use up some of the reserves, part of the fuel for supporting industry consumption beyond the year 2000 will have to come from newly proved reserves, most likely by firming up what are now called "probable" reserves. Drawing on non-U.S. uranium reserves would substantially improve the picture. Naturally, emergence of the breeder would make these calculations academic.

world shale oil resources have been estimated at some 350 times the volume of ultimately recoverable petroleum. Moreover, it is widely distributed. In the United States the richest formations are located in the Rocky Mountains, primarily in western Colorado. While estimates at best are highly approximate, the numbers are so large that even a 50 percent overstatement would allow an extraordinarily rich source of liquid energy.

There are several reasons why this energy resource has remained largely unexploited. For one, throughout the 1950s and 1960s, cheap petroleum, both domestic and imported, kept shale oil from becoming a competitive alternative. Potential producers were then saying that the cost per barrel would be 25 cents or so more to produce shale oil than petroleum. Some R&D work was undertaken, by a few oil companies, as well as by the Bureau of Mines, but there was every indication that at prevailing oil prices shale oil had no future. Shale oil was, however, the conventional fallback position, and no popular writer on energy failed to observe casually that if the United States ever were to "run out of petroleum," it could then turn to shale oil for its liquid fuel needs.

It has not turned out to be that simple. Petroleum prices have risen spectacularly, but shale oil is nowhere to be seen, even though it has for so long been "just around the corner." There must be other reasons for this, and, indeed, there are. Above all, although shale oil has been produced for more than one hundred years in Scotland, in the Baltic region, and elsewhere, a technology for large-volume production under environmentally acceptable conditions has yet to be perfected.

There are two basic ways of achieving this. In one, the parent material is mined as if it were coal, and the oil is recovered above ground through distillation in retorts. The other process is designed to short-cut the mining stage. The heart of this *in-situ* process is the extraction of the oil by lighting underground fires, which causes the oil component to liquefy, after which it is pumped to the surface.

Both the approaches have their problems. The mining-retorting method is costly, since it involves a separate step, requires large retorts, and produces residual material that—because it expands in retorting—exceeds by volume the amount of shale mined. Its subsequent disposal poses difficult environmental problems. The *in-situ* process has proved difficult to operate on a commercial scale. Finally, it so happens that oil shale is concentrated in areas of the country that are valued for their natural environment and have not so far been the site of industrial activities. These areas are also sparsely settled, and the prospect of an intrusion of a large work force, first for construction and later for operation, has encountered considerable apprehension and, at times, open hostility. Thus, oil shale development has long been embroiled in environmental controversy. By the same token, steps taken to meet the objections would add to the costs and postpone the time at which shale oil could become competitive.

A few projects, some assisted by government funding, are now in their initial stage, but as ideas are translated into engineering charts and charts turn into boilers and valves, costs have risen steadily. From the perspective of the early 1980s, it looked as if shale oil might once again be returning to a position "in the wings." Exxon's decision in early 1982 to end participation in an ambitious multibillion-dollar shale oil production project in Colorado reinforces that perspective, though a much smaller Colorado project, undertaken by Union Oil, was looking toward completion in 1983.

Tar Sands

This material, also referred to as bitumen, consists of large sand beds permeated by ancient organic materials. The trick is to disengage the hydrocarbons from the sand. Large quantities are located in Canada, the United States, and also are known to exist in Venezuela, Zaire, and Trinidad and Tobago. This list of scattered geographic locations suggests that it is not an energy source that has been sought very avidly. Estimated resources—there are no estimates of reserves, that is, amounts now producible—in this country, largely located in Utah, are sizable, in the neighborhood of 30 billion barrels.

Those in Canada are believed to be five to six times as large as those in the United States. These sands lie very close to the surface, and the oil contained in them can be recovered in a heating process. Two Canadian operations have successfully processed tar sands. After several years of losses, they can now produce profitably at world prices. Harsh local climate conditions have added to the difficulties these ventures have experienced. The intense cold of Northern Alberta has caused frequent breakdowns of some of the principal equipment. No commercial production exists elsewhere, and interest in developing the Utah deposits has not been high. The reason basically is one of cost. The concentration of oil per unit of weight or area is low, and, as compared with crude oil that emerges ready for refining, tar sands, like oil shale, require intricate treatment by machinery under heat to coax out the liquid.

Heavy Oil

Oil that is very viscous, that is, thick and sluggish, is called heavy oil. Unlike normal crude, this type of oil will not move to the wellbore and gush to the surface, nor can it be readily recovered by pumping. Other steps must be taken to make it flow. Heating it by injecting steam or burning part of the oil in place are two ways of tackling heavy oil. Some of the experience gained in enhancing the recovery factor of conventional crude may be useful in the future to exploit this oil. A great quantity exists; in the United States alone, heavy oil in place is estimated at more than 100 billion barrels, but, at best, one-fifth of that may be recoverable in time. Nearly 500,000 barrels a day are now produced in the United States, almost all of it in California.

The Venezuelan situation illustrates what is at stake elsewhere. One of the best-known formations lies along the Orinoco River. It has been estimated to range between 700 and 3,000 billion barrels. The Venezuelans believe that 500 billion barrels might be recovered in time. The pace would, however, be quite slow: at most, 1 million barrels per day by the year 2000, given the difficulties and, therefore, the cost of extraction, as well as the problem of upgrading the quality of the product, which is high in sulfur and metallic components.

UNCONVENTIONAL GAS RESOURCES

Unconventional gaseous energy sources are also very large. If they could be economically tapped, the United States would be well supplied far into the next century. No wonder they have lent themselves to wildly exaggerated reports in the media. What is usually given short shrift are the unsolved technological problems and the cost. This is not to say that these resources do not exist. Indeed they do, and some are being produced commercially. But it is unlikely that any of them will make significant contributions to the nation's gas supply during the balance of this century. Moreover, while knowledge is inadequate for U.S. resources, it is virtually nonexistent for the rest of the world.

The principal kinds of gas in this category are gas from Devonian shale, gas in tight sands, methane from coal seams, and gas in geopressured zones. *Devonian shale* takes its name from a rock formation of the Devonian period that, in North America, is found mainly in the Appalachian Basin, stretching from Michigan to Alabama. Gas deposits occur at a relatively shallow depth, and there are now thousands of wells from which gas flows to the surface. A major problem is that these flows are very slow so that annual yields from a given well are small, which leads to very slow recovery of the investment. Moreover, to produce significant quantities, it is necessary to sink a great many wells. The estimated magnitude of this gaseous material—nearly 50 percent more than proved reserves of conventional natural gas—is thus robbed of much of its significance.

A large quantity of gas also is contained in *tight* sand formations, the same kind of geologic formations that normally contain natural gas. However, in this case, the sand formations are too dense to permit the gas to move readily, if at all, to the wellbore when a well has been drilled into the gas-bearing formation. Various ways of fracturing the deposit, including underground nuclear blasts, have been tried, but so far none have been sufficiently successful to permit these resources to be included as early additions to reserves. As of mid-1978, the Department of Energy put the magnitude of this resource at nearly four times conventional natural gas reserves. Most deposits are located in the Rocky Mountain states, especially Wyoming, Colorado, and Utah.

Third, there is a great deal of methane gas in *coal seams*. To produce it would serve a dual purpose: (1) it would enhance the safety of coal mining, if the gas were removed prior to the commencement of coal mining, and (2) the gas would be an addition to the nation's energy patrimony. The amounts are large, ranging from one and one-half to three and one-half times conventional gas reserves. There has been some exploitation of this gas in Europe but not in the United States. Apart from technical issues, there is the yet-unresolved legal problem of whether ownership of the coal seam comprises ownership of the associated methane.

Finally, there is the largest potential gas resource of all, the methane dissolved in *geopressured brines*. The brines are found along the Gulf Coast of Texas and Louisiana and abroad in many locations. This methane gas is dissolved in salty waterbodies at various depths below ground, under substantial pressure and normally at elevated temperatures. This gas can be recovered by drilling into the formation to tap the brine. At the surface, the gas will separate out, leaving the hot brine to be disposed of.

This resource is truly vast—amounting to at least fifteen times conventional natural gas reserves, and possibly far greater amounts; there are, however, numerous uncertainties regarding it. One involves the fraction of the gas in place that can be recovered. This may be as low as 5 percent, but even at that the resource is substantial. Under the most pessimistic resource estimate it would almost equal natural gas reserves. Another problem is that each occurrence has its own characteristics regarding porosity and permeability. Moreover, none of the experimental wells has flowed long enough to provide an understanding of the behavior of the resource over time, but there are indications that it will vary from site to site. Next, there are important environmental issues, including subsidence of the land as the brine is pumped out, the disposal of the brine itself, and others. Costs appear substantially higher than prices of even those categories of natural gas not controlled under the Natural Gas Policy Act.

All told, the frequently heard notion that here is a resource, close to the surface, in known locations, obtainable by pushing a hole into the ground, and so vast that it will solve the nation's energy problem is wildly off the mark. A great deal of patient research remains to be done, and it cannot be taken for granted that commerical exploitation will ever become feasible, or if so, there is no way of predicting when this might come about. Even though geopressured reservoirs are spread widely around the globe, gas has not been produced anywhere, nor are there even highly approximate estimates of this resource's dimensions.

With the decontrol of prices for much of the nation's natural gas approaching and a substantial boost in price inevitable, the time when these various unconventional gas sources become competitive will move closer.

Unfortunately, the lack of trustworthy information on the anticipated costs of these gases leaves matters in limbo.

RENEWABLE ENERGY RESOURCES

In a strange twist of history, the energy sources that humans have relied on longest—water, firewood, the sun—and from whose limitations the transition to the fossil fuels has freed us now are looked upon as highly desirable. What has brought about this reversal? For a reversal it surely is. As noted in Chapter 1, the energy sources relied upon by our ancestors had their drawbacks—low levels of concentration and poor efficiency; unreliability of supply, and, in general, lack of control; and limited ways of turning heat into other forms of energy, such as kinetic or mechanical. The enormous amounts of potential energy in coal, oil, and gas changed all that. Modern energy sources fueled the industrial and transportation revolution. The renewable—or nondepletable—resources were largely abandoned or reserved for special occasions and applications, except, of course, for water, which was harnessed for electricity generation, and is used so to this day.

What then brought the renewables back into center ring? One factor was the rapid rise in demand that portended a much earlier depletion of at least low-cost oil and gas resources than formerly had been anticipated. A second was the environmental effect of fossil fuel use. A third is the widespread perception that renewable resources are more benign in the social sense than nonrenewables. A special term has been coined for this phenomenon, *the soft path,* suggesting that the use of renewables not only has environmental advantages, but also could lead to a deemphasis of the giant industrial facilities and organizations now mainly responsible for energy supply and to their subsequent replacement by small-scale, community-sized and -oriented supply centers. In its present form this approach also envisages politically benign consequences of such transformation, viewing it as a better environment for a democratic society.

One need not agree wholeheartedly with these speculations, and some appear rather fanciful, to see the main advantages of the renewables—they are nondepletable and avoid some of the most disconcerting forms of pollution. Given the large potential for increases in energy demand—emanating from especially those countries who today use only a fraction of energy per capita or per dollar of goods and services consumed by the industrial nations of the world—a turn toward sources other than the fossil fuels is a highly desirable goal.

Nuclear power is an energy source, especially in the form of the breeder and of fusion, that also offers the vision of near inexhaustability (see Chapter 4). But the breeder, embroiled in controversy, is at best two or three decades away from making a dent in the energy picture. And we have no real assurance that fusion will become a commerical energy source. In a balanced view of the future it is well, nonetheless, not to lose sight of the potential role of the breeder and of fusion as candidates, alongside the renewables.

In view of the patent advantages that a return to renewables would have, and the dominant position of the conventional energy sources that severely constrains the speed with which a mix in the energy pattern can be accomplished, it is not obvious why the debate on the role of the renewables has become highly emotional. This probably stems from underlying perceptions having little to do with energy that find energy policy a fertile ground for testing one's convictions. In what follows we shall try to strip the topic of its emotional overtones, persuaded that what one has to say about renewable energy stamps one as neither morally "good" or "bad" nor politically "wise" or "foolish," but rather as aware or oblivious of both the promise and the problems.

The renewable energy sources that have received most attention in recent years are biomass and solar. *Biomass* is a generic term encompassing such organic substances as trees, annual crops, organic wastes, and so forth. The term *solar* is employed narrowly to refer to the direct utilization of solar rays; others use it more broadly to include energy sources that are indirectly associated with solar radiation, such as falling water, trees and crops, ocean waves and temperature and salinity differentials. At times the concept is stretched to encompass also geothermal. It is solar energy in this all-inclusive definition, incidentally, that some believe can meet as much as 20 percent of total U.S. energy consumption by the end of the century. Presently it accounts for about 5 or 6 percent; a precise figure is not available because data on wood use are poor.

In sharp contrast to inquiries into fossil fuels, the least significant and interesting resource question is, How much is there? The answer generally is, "An awful lot." This holds not only for solar, but for all renewable sources. The three critical questions that an inquiry into renewable energy sources must address are:

- At what cost can the energy be produced and utilized by the end user?
- How soon can the various sources and modes of use become significant contributors?
- Can the renewable energy sources make headway without substantial assistance from government?

It means also that in discussing these sources we are drawn unavoidably into the area of technology. Thus, what follows will anticipate to some extent the scope of Chapter 4 and should be read in conjunction with it.

Because solar radiation is pervasive and nondepletable, it makes little sense to incorporate it in the conventional resource terminology. Limitations on its use are not lack of availability, except at night and on cloudy days. But its accessibility—that is, having it in the right space at the right place—and by the cost of materials—metal, glass, plastics, and so on—for heat collectors, and silicon or other substances for solar cells. Thus the issue of adequacy, a critical consideration for fossil energy sources, has a different meaning for solar energy. Measured as total solar radiation per year per given unit of area, it denotes the resource available in the broadest sense. But unlike oil and gas there are no "reserves" indicating the amount of energy to be recovered under current technological and economic conditions. Rather, there are various systems of capture, each carrying different installation and operating costs.

The first important question to ask is, Are they competitive with conventional sources? Even those that can compete economically may not come rapidly into widespread use, either because (1) they are novel, and potential users will stick with more conventional sources; (2) backup supplies may be required for sunless periods; or (3) their integration into utility networks may pose problems. To overcome the reluctance of potential users to switch to solar energy, a number of programs have been designed for government-assistance, just as in the early days of industrialization the government aided so-called infant industries. And that becomes the third important issue. In brief, solar energy is virtually unlimited, nondepletable, and can be captured by anyone. However, capturing it has its cost; both space and materials have a cost and are finite in quantity, and institutional constraints may be hard to overcome rapidly without public policies that ease them.

It is convenient to differentiate at least three categories of solar technology narrowly defined: solar heating and cooling; solar thermal; and photovoltaics. The first classification in turn may be subdivided into passive and active, depending on whether solar radiation is merely given a better chance to do its work—through appropriate location, construction techniques (for example, use of insulation), and operating practices in a building, or whether equipment (such as pipes, pumps, fans, or solar collectors) is employed to collect and concentrate, distribute, or store the heat more efficiently. Both passive and active solar systems are viable today. Appropriate material and equipment is commercially available, and firms can be found to make the necessary installations and provide services. Solar water heaters are now advertised in some catalogs, along with gas and electric ones.

The question regarding heating and—less well developed—cooling is not then whether it is feasible or available. It is both. Rather we should ask, What is its cost? Are there technical problems? How quickly might such technology be adopted? Answers to these questions must be tentative at present. For new buildings, the initial costs are significantly higher, but these may be recovered over time, even allowing for auxiliary conventional energy supplies to back up the solar-driven system. Enough experience is not available, however, to determine how long that payback period might be. And since Americans frequently move, the initial high cost is a drawback for those who are more mobile. In

Finite and Renewable Resources

It always has been known that some natural resources—such as fossil fuels—are finite and can be depleted, while others—such as air—are nondepletable. Even some, however, that are basically nondepletable, can be degraded in quality, as we have learned traumatically in the past two decades when "good" air and water have become scarce resources.

Renewability is still another matter. Crops and trees are renewable, even though poor management can make them nonrenewable for a given time and space. Solar energy is termed renewable only by stretching the word; it is, however, nondepletable. Indeed, solar energy is perhaps the first energy form that is wholly nondepletable. No degree of use can diminish its magnitude. What is depletable are the materials employed in making solar energy useful, and what is finite is the space in which it can be captured.

Speed and degree of depletion is largely a matter of cost. There are no sharp dividing lines. For example, Vermont granite and Chattanooga shale contain vast amounts of uranium, but extraction would presently incur exorbitant costs. New technology could reduce these costs and thus greatly alter the degree of finiteness. Nonetheless, there remains a difference in principle between renewable resources, even though degradable, and depletable resources such as oil or coal or uranium, even though extendable through science and technology.

time, however, one would expect this factor to diminish as experience accumulates. As for the speed of market penetration, that will depend heavily on the rate at which new construction occurs. Turning existing buildings into solar-heated ones is difficult and costly at best and, at worst, impossible. Older buildings have not been constructed to take full advantage of sunlight; in fact, often the opposite is true. They cannot be swung around to achieve a better exposure or have their walls modified to admit or exclude sunlight. New housing is the key, and while needs are large and growing, the housing industry has been for some time in poor shape. Its vigorous revival is a precondition for the growth of solar heating.

A second direct use of solar radiation is its conversion to heat. This either can be used directly or indirectly to generate electricity. Various schemes are under way using concentrating devices—troughs, dishes, bowls, and so on—some tracking the sun and focusing either on nearby pipes carrying fluid or on a single point where the combined reflected radiation produces steam. The area in question can be quite large, and installations of this kind are likely to be practical only in areas of almost continuous sunshine, so that the capital investment can be spread out over the maximum period of solar irradiation. The experimental phase is in full swing, but it is too soon to anticipate the conditions under which this technology may be competitive with conventional steam-based power generation. Plates 2 and 3 show such a power tower and the associated mirrors in Barstow, California, that began generating electricity in November 1982. Similar experiments are being conducted in France, Italy, Japan, and elsewhere.

A third area of application is that of photovoltaics. It was 1958 when solar cells passed their practical test by powering a radio transmitter in a U.S. satellite. (See Chapter 4 for a discussion of the current status of photovoltaics.) Again, the issue of achieving greater efficiency is the heart of the matter. Several companies today offer photovoltaic cell systems, and industry has made impressive strides toward making a commerical success of the principle that passing sunlight through certain materials can excite an electric current.

There are important pros and cons. The technology constitutes the most direct conversion of sunlight into electricity. It involves no steam, turbines, generators, and the like; thus, operation and maintenance costs should be low. Second, the system is modular, that is, it can be built up to any size by adding extra cells. This enables it to meet needs of different magnitudes, a characteristic especially important at the lower end of the size range where conventional technologies tend to become inefficient, costly, or both. On the other hand, of course, no economies of scale can be reaped. Third, it can operate in any environment, not in need of special conditions (other than sunlight), cooling water, and so on, and it does not pollute ambient air or water. Its space requirements are substantial but not overwhelming (see Plate 4).

However, two problems remain to be solved: the cells that are the operating part of the system are expensive and inefficient, that is, they convert only a small portion of the sun's rays that strike the cell into electricity. Thus, the search is on for both a cheaper material and lower costs of preparing it, and for ways of increasing the conversion efficiency. Optimum efficiency of the best single cell today is still less than 20 percent. That efficiency drops to less than 10 percent when mass-produced cells are finally assembled in so-called arrays, that is, frames in which large numbers of single cells are situated. As for costs, both the cells and, perhaps even more critically, the support system must become cheaper. The latter point is often overlooked. Yet, it has been said that even if the cells were available at no cost, the system in which they are incorporated is today too costly to be competitive. Thus, there is a continuing search for cheaper support materials.

Costs have diminished sharply during the past two decades; they are about one-twentieth of what they were in 1959. This has given rise to hopes that there is a large reservoir of ingenuity in the photovoltaic community (which now includes several large oil companies) virtually to guarantee a continuation of that process. That is probably too sanguine a view, but there is little doubt that this solar technology has made important strides and that in time it may become a true competitor in the generation of electric power. That the utility industry is carrying on substantial research in this field—especially by way of the Electric Power Research Institute—supports this hypothesis. According to EPRI's annual survey, in 1981 the 234 utilities responding were engaged in 943 solar projects, only half of which dealt with heating and cooling, as contrasted with 1975, when 83 percent were in that area (Figure 3-3).

Given the limitations of this volume, it is not feasible to deal equally extensively with each of the remaining renewable resources: biomass, hydro, wind, geothermal. Nor is it necessary. *Hydroelectricity* is a tried and true technology. Currently, attention is not focused on building more large dams. There are not many sites left, except a few in the developing countries, such as those along the Zaire (Congo) River; and in the United States, at least, the destruction of scenery, including the river itself, is bound to lead to bitter and costly controversy. On the other hand, there exist small sites, some of which, though now abandoned, were formerly active and could be—and are being—reactivated. It is unlikely that in the aggregate they will make a major contribution to the national supply of electricity, but they do serve local needs.

Another time-honored energy source is the *wind,* but modern wind turbines are a far cry from the old windmill still found on some farms. The latest experimental models are 200 feet high, and their rotor has a diameter of 300 feet (Plate 5 shows an earlier version). Models now on the drawing board will be even higher; the diameter of their rotor blade will be 400 feet, which is one-third longer than a football field. The entire structure would weigh some 600 tons.

These machines have advantages and disadvantages. Among the pluses are possible low costs per unit of installed capacity, once the technology has been perfected; the ease of introducing them piecemeal, as is the case for pho-

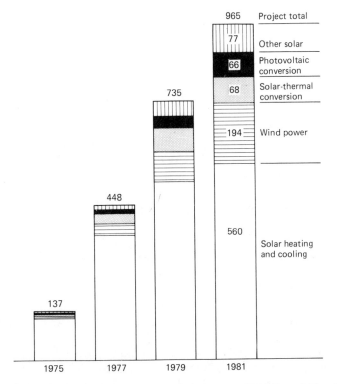

Figure 3-3 Utility industry solar energy projects. [*Source: EPRI Journal,* Electrical Power Research Institute, Palo Alto, California (December 1981).]

tovoltaics; and relatively little demand on land in the sense of direct denial of land-based activities. One disadvantage is their discontinuous operation except in locations where the wind blows incessantly, or almost so. Another is that winds below 15-mph and above 45-mph are not acceptable. That is to say, the machine will be shut down under those conditions.

Again, it is idle to speculate how soon and how much this technology might contribute. We are witnessing the early development stages, which are in competition with other established, as well as novel, technologies. Nonetheless, the electric utilities have evinced substantial interest in its development, envisioning it as a possible intermittent replacement for fossil fuels in appropriate locations.

Like wind, *geothermal energy* is both old and new. Hot steam and briny water belching from the earth long have been known. One such source, still in operation today, was used nearly eighty years ago to produce electricity in Italy. More recent installations are located in California, New Zealand, Iceland, Japan, and Mexico. All told, their significance so far is small; but the existence of many promising locations elsewhere in the world suggests that geothermal could become an important energy source, obtainable either by tapping underground steam, which is preferable but much rarer, or by tapping underground hot water, a widespread but technologically less attractive technique. The dissolved minerals in the hot brine are corrosive, they impede water flow, impair the transfer of heat to the generating components, and their ultimate disposal presents problems for the environment.

Another approach that has received much attention during the past decade is so-called dry-rock geothermal. This technology requires the drilling of a well to reach the hot magma in the earth's crust. Water is sent down the shaft, and as it heats it is drawn off through a parallel well and is used to drive a turbine-generator. This is an expensive technology. Deep wells are required to reach the hot dry rock and two wells, rather than one, must be drilled, usually into very hard rock. How long each well can remain active is a yet-unresolved question. Again, efficiency and cost will be the ruling variables. Between the different forms of geothermal it is likely that those focusing on existing underground hot-water sources will be the most successful for some time to come. Their combined potential is awesome, so are the obstacles to overcome.

What about *biomass*? While rightly considered renewable, biomass is, in a special sense, "depletable" in that space to grow is both physically limited and economically contested by alternative uses of the soil and the crops themselves. Although the reserve-resource terminology lends itself poorly to expressing the potential value of biomass as an alternative energy source, it has not kept people from using it. For example, it is calculated that if the entire corn crop of the United States were dedicated to producing alcohol for use as liquid fuel, the amount of such energy would be in the neighborhood of 450 million barrels, or equivalent to only 8 percent of current U.S. liquid fuel consumption. Similar calculations can be made for land areas dedicated to trees of given characteristics and period of growth. It is of interest to delineate the dimensions of the biomass contribution, but unless the limitations of such measurement are recognized, this does not otherwise serve a useful purpose, since the extent to which these sources will enter the consumption stream will be shaped by factors other than the resource base.

Recently, energy experts have studied two greatly different sources. One is direct wood-burning. The U.S. Department of Energy estimates that wood-burning, in individual stoves and, more important, in wood-associated industries, contributes about 3 percent to the U.S. energy supply. Increases are likely, especially in residential use, but inconvenience as well as emerging air pollution problems are significant impediments. In much of the developing world, of course, fuelwood plays an enormous role, and has a substantial impact on land use and many unsolved issues plaguing the entire forest economy.

The more recent trend is toward the conversion of biomass to a liquid form, for example, ethanol or methanol. For the reasons given above, crops grown on valuable cropland generally are not used, except for sugarcane, which has been successfully used as feedstock for synthetic gasoline production in Brazil. On the other hand, trees grown specifically for harvesting as methanol feedstock are a better bet. Their potential contribution is much less limited by the disproportionate magnitudes of liquid fuel demand and the area required to meet it. Roughly, one could derive about 40 barrels of liquid (mostly methanol) from 1 acre of trees. To meet, say, 10 percent of U.S. liquid fuel consumption in that way would call for about 15 million acres in trees, a large but not fanciful magnitude.

TURNING RESOURCES INTO SUPPLY

The fact that reserves and resources exist does not mean that it pays individuals or society to make use of them; nor can estimating their magnitude lead one to predict the annual flow of supplies. Too many factors influence what will actually be produced. Within two years, for example, exports by the OPEC nations declined from just under 30 million barrels a day in early 1980 to about half that level in early 1983. In this case, economics—not changes in the physical existence of oil—choked off the flow. Reserve and resource estimates convey a sense of what might be the sustainable limits to producibility, but are of little help in estimating near-term supply flows. Forecasts have fluctuated throughout the seventies, as forecasters make their own assumptions about costs, prices, and markets.

Over the past decade, until about mid-1981, most prognosticators had assumed that, inflation apart, the price of energy would rise at an annual rate of 2 to 3 percent. This increase, they argued, would follow from several factors:

- Continued increase in worldwide demand, even though it would fall short of the rate experienced in the 1950s and 1960s.
- Higher resource costs which would stem from the more adverse conditions under which future exploration for oil and gas would be undertaken. For example, the necessity for drilling at ever-greater depths, in offshore locations, under unfavorable climate conditions such as those in the Arctic, and at greater distances from existing transportation systems
- The public's heightened interest in preserving the environment would increase clean-up costs associated with exploration and extraction.
- Finally, the actions of governments here and abroad to deregulate oil and gas prices, and OPEC's export policies.

Less-affected energy sources such as coal could see their prices increased as they became partial or full substitutes for oil. Nevertheless, it would take a while before cost pressures on coal, now having ample excess capacity, would begin to operate. Looked at differently, the steadily but gently rising price level would constitute the incentive for calling forth needed supplies by enabling producers to recoup the rising cost of producing them. At the same time, according to this scenario, oil would gradually give way to coal and nuclear energy in uses where it could function as a substitute, such as providing steam for electric power generation or heat and power for industrial processes.

This prospect was widely accepted, and future flows of conventional energy were mapped over ten to twenty years by research bodies, the federal government, and the energy industry. Usually these projections showed a declining use of oil and natural gas, a rising use of coal and nuclear, and a small, but rising infusion of nonconventional sources. At the onset of the 1980s, this conventional outlook picture was modified substantially. Energy consumption had dropped much more dramatically than had been anticipated, as a consequence of the depressed state of the world's economies and of the reduction in consumption, especially of oil, in response to the second round of OPEC price boosts in late 1979 and early 1980. The big question that emerged was whether the high OPEC price would be sustained by drastic cutbacks in production, or whether it would be forced down by a sustained decline in consumption. If production were cut back, oil would continue to be phased out as the leading energy source more rapidly than had been anticipated in the seventies, but it would last for a greater number of years. If prices were to fall, the sudden and steep decline in oil consumption could be reversed as alternative energy sources lost attractiveness when assessed in the light of lower oil prices. The degree to which the steam seems to have gone out of the U.S. synfuels program bears witness to this possibility. The return to larger, less energy-efficient motor vehicles would be another signal.

Most projections continue to assume a rapid decline of oil's share in total energy use, offset by increases in coal and nuclear; but at the same time most projections also assume a substantially lower growth of total energy consumption (see the discussion in Chapter 2). The net result is that the outlook foreshadows fewer and less painful price and supply problems than was true in the seventies. The cutback in price and output decided by OPEC in March 1983 has given an initial answer to these questions, but the longer-run implications are far from clear.

A general easing may be in the offing, but one cannot bank on it. Why? Above all, because the world oil supply framework is very fragile—not everywhere, but certainly, and most of all, in the critical Middle East. With more than half the world's oil reserves located in that part of the globe, any major, sustained disturbance would have an immediate impact on prices, as happened in 1973–74 and again in 1979–80, or as might be the case if loading facilities in the Persian Gulf were to be destroyed or its waters made inaccessible to shipping. In such an event, the world would quickly draw down its accumulated stocks of crude and refined products, prices would skyrocket, and the economies of the oil-importing countries would once again come under stress. Moreover, to the extent that the renewed abundance of oil has discouraged the development of alternative liquid energy sources, one could expect little, if any, relief, should oil once again become scarce and expensive. While it was never realistic to expect daily production of 500,000 barrels of synthetic liquids by 1987, let alone 2 million barrels by 1992, as specified in the Energy Security Act of 1980, it is quite clear that the drive to build a synfuels capacity has lost momentum and that it will be much later than 1987 before any significant production emerges.

Nor is this all. Optimistic predictions of a return to "normal" are based on a continuing rapid increase in the use of both coal and nuclear energy. These are tenuous assumptions at best.

On geologic and economic grounds alone, coal production could increase by several percentage points a year for many years to come, with western coal responsible for a growing share of the total. And if the opposition to coal pipelines (see Chapter 1) could be overcome, even the transportation of coal would raise less concern. But while the adverse environmental consequences of coal production and combustion have not prevented a fair rate of growth, especially as contrasted with coal's performance in the 1950s and 1960s, they will continue to have a dampening effect. This is especially true of industrial users who have many reasons for staying away from coal. Since 1973, coal use by utilities has grown by nearly 50 percent. In sharp contrast, industrial use (that is, excluding coking) which accounts for not quite one of each 10 tons of coal consumed has declined by some 10 percent during the same period.

Of growing concern is the generation of acid rain which could well become a major threat to increased use of coal, provided that ongoing research efforts confirm that coal combustion is a major contributor to this phenomenon as well as costly to control.

As for nuclear power, whose prospective rapid expansion played a significant part in earlier predictions of coming ease at the beginning of the eighties, the uncertainties are great and many. It would be foolhardy, indeed, to take for granted that all construction plans now on paper in the United States and the rest of the world will come to be executed. Again, it is not a question of an adequate uranium supply. There is more than enough for current projections of nuclear power growth. Instead, there are a number of other factors, most of them closely interrelated, that must be considered:

- A large number of the public continues to be uncomfortable with this particular source of energy whose origins inescapably link it to the bomb.
- The exceedingly serious, widespread, and long-lasting consequences of a major mishap

- A utility industry that until recently has looked at nuclear fission merely as a different way of raising steam under a boiler
- The possibility, however remote, that somehow nuclear material could be diverted to the production of explosive devices in the hands of either unstable nations or subversive groups
- The interminable delays in fashioning a system that can be depended upon to render nuclear waste innocuous
- A long-standing habit of government to make light of the darker side of the technology, thus creating distrust of official pronouncements when things do go wrong.

All of these doubts and issues add weight to the sinister side—or as nuclear scientist Alvin Weinberg put it, the "spooky" aspect—of nuclear energy. Incident after incident—none of them producing casualties—appears to confirm the unfavorable image. Despite the fact that taken together they have written an infinitely better safety record than, say, coal mining, each one raises anew the question, What if..., and is followed by the specter of widespread death, disease, and devastation.

Regulations are written to enhance safety and do so. But at the same time they raise the cost—either by extending the time it takes to move a new plant from blueprint to its first day of operation, or by unceasing needs to add new equipment, instrumentation, and so on. In an era of high-cost money, this can fatally reduce competitiveness. Utilities, being under no great pressure to expand capacity, have responded in the last five years by canceling and postponing the construction of nuclear power plants. Nor can managers be indifferent to public opinion: even a minority, when active and vocal, can be a powerful deterrent and indirectly contribute to already rising costs.

By mid-1982, installed nuclear power capacity was just over 50,000 MW. No new nuclear plant has been ordered in the United States since 1978; and sixty-two plants were canceled between 1974 and 1981, and more in 1982. As against an earlier projection of 734,000 MW by the year 2000, few would now quarrel with the notion that, at most, 165,000 MW will be in operation by that time. Even that low figure rests on the highly questionable assumption that all plants on order or under construction in 1982 in fact will be completed. Thus, some believe that a capacity of 100,000 MW to 120,000 MW is a more realistic guess. What does such an estimate imply? To be sure, the growth rate of 7 percent per year in electric power demand that prevailed until 1973 is a thing of the past. It is more likely to be 2 to 3 percent now, and electricity supply margins seem adequate through the 1980s, provided that (1) aging plants are kept in good repair to prevent increasing outages, (2) demand does not unexpectedly pick up, and (3) new generating facilities of all types now on company books indeed are constructed on schedule. But within ten or fifteen years, added capacity will be needed, and, given the long lead times that show no sign of shortening, new plants must be started soon. Unless the nation has by then come to terms with nuclear energy, these plants will be coal-fired, and that prospect, implying perhaps twice the coal use that exists now, will bring to the fore the question whether the world is approaching environmentally dictated limits to the burning of fossil fuels and especially of coal.

To summarize, the supply outlook is reassuring only if a number of underlying assumptions regarding oil, coal, and nuclear turn out to be realistic. There is legitimate disagreement on this, and that alone is reason enough for keeping alert to the possibility of trouble.

ABOUT "RUNNING OUT" OF ENERGY

The discussion of resources and reserves, the outlook for supply, and the importance of the underlying assumptions will have made it clear that it makes little sense to speak of the country or the world as "running out" of energy. When the supply flow thins, prices increase, consumption declines, and, aided by new technology, the supply of the commodity in question, or a workable substitute, increases. A new equilibrium emerges at which supply and demand are brought into balance at a higher price.

Thus, the world will not run out of oil, but it will have to pay more for it. As oil becomes more costly to find, extract, and transport from remote places, its use will decline. Customers will abandon it for other sources of energy, or use less energy altogether by substituting other goods (blankets, insulation), or both. Oil will continue to be used by those who value it enough at the higher cost not to go without. It pays them to buy at the higher price. If the market works—a proposition one should not take for granted—the transition should not be too painful and petroleum will be used efficiently, that is, in uses where it has the highest value to the user.

This may seem unrealistic or "academic" to many readers. After all, the United States is "running out" of oil, in the sense that each year less and less is found and consumption outpaces additions to reserves. One easily can produce calculations that show the year in which it will all be gone, but these calculations are intended merely to convey an idea of the degree of abundance or scarcity of the material in question. They are not meant to draw a time profile of future consumption, as many tend to interpret such exercises. The world will "run out" some day in the sense that it will become too costly to use oil, when compared to the cost of substitutes. But it is altogether likely that at that time there will be much oil left in the ground, for instance, much of the two-thirds not pumped out during past production. Distinguishing the physical from the economic world may at times seem artificial, awkward, or both, but not doing so is a real barrier to understanding the supply side of the energy equation.

BIBLIOGRAPHICAL NOTE

There is an extensive literature dealing with both quantitative and conceptual aspects of energy resource estimation. The references cited for the tables in this chapter are leading sources of data and information on resources and reserves.

None of them can be said to be *the* authentic source. High among those worth consulting are the estimates compiled by the World Energy Conference, contained in *World Energy Resources 1985-2020* (New York, IPC Science and Technology Press, 1978), the wide-ranging effort, *Energy in a Finite World: Paths to a Sustainable Future,* Report by the Energy Systems Group of the International Institute for Applied Systems Analysis (Cambridge, Mass., Ballinger, 1981), particularly chap. 3, and Joseph D. Parent's *A Survey of United States and Total World Production, Proved Reserves, and Remaining Recoverable Resources of Fossil Fuels and Uranium* (Chicago, Ill., Institute of Gas Technology, 1982). The last is outstanding for its careful explanations and discussion of the nature of the data. Updated versions of both are likely to appear subsequent to the above-cited edition. Oil and gas resource and production data can also be gleaned from various issues of the weekly *Oil and Gas Journal.* Worldwide data on uranium are compiled by the Organisation for Economic Co-operation and Development in Paris (OECD) and published in *Uranium: Resources, Production and Demand*, which is periodically updated.

The primary sources in the United States are the U.S. Geological Survey (USGS) and the U.S. Department of Energy. On oil and gas the USGS published its basic resource estimates in 1975 under the title of *Geological Estimates of Undiscovered Recoverable Oil and Gas Resources in the United States,* Geological Survey Circular 725. An updated version was released in 1982. Until the emergence of the Department of Energy, coal resource estimates were computed and published by the U.S. Bureau of Mines. They are now published by the U.S. Department of Energy, found most conveniently in the *Annual Report to Congress* by the Department's Energy Information Administration.

To understand the intricacies involved in making resource estimates, one of the most useful discussions rather than data can be found in "Oil and Gas Resources—Welcome to Uncertainty," [*Resources* (March 1978), Resources for the Future, Washington, D.C.]. It in turn carries a number of references to other writings on the subject. Attention is called also to Richard Nehring's *The Discovery of Significant Oil and Gas Fields in the United States* (Santa Monica, Calif., Rand Corporation, 1981), followed more recently by the same author's "Prospects for Conventional World Oil Resources," *Annual Review of Energy,* vol. 7 (1982). A brief but comprehensive discussion can be found in *Energy: The Next Twenty Years* (Cambridge, Mass., Ballinger, 1980).

Because of its role in judging the need and timing of breeder reactors, there was a heated debate on the magnitude of uranium resources in the mid- and late seventies. To get the flavor, the reader is referred to *Nuclear Power: Issues and Choices* (Cambridge, Mass., Ballinger, 1977) and to *Energy in Transition, 1985-2010,* Final Report of the Committee on Nuclear and Alternative Energy Systems (Washington, D.C., National Research Council, National Academy of Sciences, 1979).

Regarding renewable resources across the board, a useful source is the *Report of the United Nations Conference on New and Renewable Sources of Energy* (New

York, United Nations, 1981), which in a little more than 100 pages conveys the substance of the UN Conference on that subject held in Nairobi in August 1981.

Readers wishing to probe the theoretical foundations of resource estimation may be interested in Douglas R. Bohi and Michael A. Toman, "Understanding Nonrenewable Resource Supply Behavior," *Science* (Feb. 25, 1983).

Finally, given the overlap between resource and R&D topics, the bibliographical note at the end of Chapter 4 might usefully be consulted as well.

Plate 1 Triumph Hill oil field in Pennsylvania at the turn of the century. Note the closeness of the drilling sites. (Photo courtesy of the Library of Congress.)

Plates 2 and 3 Solar power tower, Barstow, California. Each mirror measures 23 square feet. The height of the tower is 310 feet. Located on a 130-acre site, it is designed to produce 10 MW of electricity, or about 1 percent of a standard nuclear plant's capacity. Below, a close-up of the mirrors is shown. (Photos courtesy of the U.S. Department of Energy.)

Plate 4 Solar photovoltaic arrays providing power for the Oklahoma Center for Sciences and Arts. (Photo courtesy of Science Applications, Inc.)

Plate 5 A wind turbine generator on Block Island, Rhode Island, which is capable of producing 200 kW of electricity (enough to power about 50 typical homes) in an 18-mile-per-hour wind. Installed by the U.S. Department of Energy and the National Aeronautics and Space Administration (NASA), it spans 125 feet from tip to tip. (Photo courtesy of NASA.)

Plate 6 Typical surface coal mining operation. The giant "dragline," operated by a subsidiary of Ashland Oil, dwarfs cars in the foreground. (Photo courtesy of the American Petroleum Institute.)

Plate 7 Oil spill (the light area in the center) from blowout of Mexican well in the Bay of Campeche in the fall of 1979. It took nine months to cap the burning well. (Photo courtesy of the American Petroleum Institute.)

Plate 8 Preparing Gulf Coast salt caverns for Strategic Petroleum Reserve storage. (Photo courtesy of the U.S. Department of Energy.)

CHAPTER 4

Research
and Development:
Widening the Energy
Horizon

Research and Development (R&D) is not a recent phenomenon. Only the name is new; the substance is old. Alessandro Volta, James Watt, or Thomas Edison would have been surprised to be told that they were in the "R&D business," but, of course, they were. In fact, Edison was in the RDD&D business, the extra two *D*'s standing for Demonstration and Deployment. In their lifetime such people were generally called inventors. Today, there are not only multitudes of individuals who engage in R&D, in small or large commercial groups, in government laboratories, or in universities, but there is also a substantial literature on just what R&D is and how it functions. In this chapter we look at the role R&D has played and can play in widening the energy options available to both this country and the rest of the world, we try to identify current trends and where they might lead us, and we ask what policies are either in place or are desirable to draw the greatest benefit from R&D endeavors.

There are various ways of looking at R&D. All of them raise intriguing issues about our institutions, public policy, international competitiveness of the goods and services based on R&D, and other matters. But a clearer understanding of the term and what it stands for is a necessary prerequisite for an intelligent assessment.

There is, to begin with, a fundamental difference between research and development. Not too long ago, research conjured up the scientist in the white lab coat, confronted with a multitude of flasks, instruments, scales, and other tools. At an even earlier point in history, it was the astronomer with a telescope peering at the stars, or Galileo watching the movement of his pendulum. It was Watt studying the behavior of steam. Now the image is harder to catch. It ranges from a solitary young scientist operating a computer keyboard and watching the TV-like screen to a large group of people engaged in building a huge device to manipulate components of atoms. Still, such undertakings continue to have a common purpose, that of improving our knowledge, without necessarily having any immediate application in mind. That is to say, although some research, and perhaps much of it, is motivated by the desire to fill specific gaps in knowledge that act as barriers to eventual development of a useful application, most research is propelled by the same challenge that makes people climb Mount Everest: "because it's there." In its purest form, research finds its home in the universities, the seats of learning.

But research is not as homogeneous as the above description implies. There is *pure or basic research,* that is, research for its own sake, the pursuing of ideas, hypotheses, and hunches wherever they might lead, just to push back the barriers of ignorance. There is also *applied research.* In one form it stops short of fashioning a useful product or service. It asks how newly gained knowledge can be employed to meet some human need. In a sense, it is an answer looking for a question; and its home is more likely to be the industrial laboratory or workbench than the university, though some recent developments, especially in the area of biotechnology, have begun to blur the distinction between the two. Sometimes, too, the research for a practical solution will lead to the discovery of the same new insight or knowledge that is the goal of basic research. At the end of the range, there is the kind of applied research that is highly empirical and pragmatic in nature and that is pursued mostly by industry. It is focused on product improvement and testing (for example, the research done by pharmaceutical companies in the search for a new or improved drug to cope with a specific disease), or new methods of quality control, and so on. It is less open-ended research, but much of it still falls into the research area though it begins to verge on what is generally regarded as a separate stage, *development.*

Some have found it useful to distinguish among three different phases: *development,* as described; *demonstration,* that is, showing on a small scale that the new product or approach does what it is supposed to do; and *deployment,* that is, the introduction through marketing, into the economy, often referred to as technology transfer.

Three important distinctions can be made among these phases. For one, costs tend to rise, often steeply, as projects move from development to deployment. Second, the degree of commitment to a selected path grows. And, third, generally the participants' roles change with each phase. For instance, as one moves along the spectrum from pure or basic research to demonstration and deployment, industry's participation increases and becomes indispensable, while government's role both diminishes and changes in its objectives. Whatever the government's reasons for supporting research might be, as commercialization is approached, it is marketability, not a governmental judgment that spells success or failure, unless, of course, government is itself the market. Similarly, while at the development stage there are still choices between alternatives, those choices vanish rapidly and tend to disappear by the deployment stage. These, then, are not trivial or capricious distinctions. They are, for example, at the center of much of the protracted debate over the future of the synthetic fuels industry—that is, who is to do what?

But even if one wants to distinguish among various phases or stages of development, they have one thing in common: they all are product-oriented. At the end of the line is a useful artifact or service that meets an identified need, that must be marketable and profit-making if it is to be developed in the private sector. In the public sector, it must meet a social need or challenge (a new medical device or treatment, a new way of coping with an environmental threat, or a new weapon or weapons system). Development is the vital link between the researcher's notebook or computer printout and the assembly line. It takes an idea, a concept, a property of matter, and moves it from theory to practice.

What distinguishes modern R&D from science and technology as it was practiced in earlier times is that it has come to be identified as a separate activity, to be funded, organized, directed like any other business. Today people "are in R&D." Route 128 outside Boston, or Route I-270 outside Washington, D.C., or Menlo Park in California are locations where R&D companies have their headquarters. The issues are: Who sets the R&D agenda? What are, or should be, the priorities? What are the respective roles of the public and the private sector? How do national efforts mesh with international ones? Assessment of these issues is helped by some familiarity with the magnitude of R&D efforts in the United States in the early 1980s, and of energy R&D in particular. Government's share, which is discussed first, is the most easily defined and identified. Industrial R&D is far more varied, and often encompassed in broader categories.

THE FINANCIAL DIMENSIONS OF R&D

In Fiscal Year 1982 the Congress appropriated about $40 billion for science and technology activity, or some $6 out of every $100 budgeted. As yet nobody has developed a test that shows whether government is budgeting too much or too little for research and development. Our only yardstick is a conceptual one—judging how much the extra or marginal dollar spent on R&D will return in the future, properly discounted to the present, and whether the dollar could be spent on something that would give a larger return. It sounds impressive, but it happens to be one of those cases where the little boxes set up to help in the calculations remain sadly empty. Thus, interested parties fall back on judging each year's budget by that of the preceding year, and by the extent to which it provides generously for their favorite project or program. Most, of course, judge that it does not.

Of the roughly $40 billion, about half are funds spent by the Department of Defense (exclusive of the $5 billion or so set up for R&D on weapons programs lodged in the Department of Energy). Space activities account for nearly another $7 billion. Health R&D takes $4 billion, and the National Science Foundation $1 billion. Within this framework for government spending on science and technology, the government's energy R&D budget comes in for at least $3.5 to $4 billion, depending on the inclusion or exclusion of some special items. We say "at least," because energy-associated R&D is funded also by such agencies as the Environmental Protection Agency, the Department of Health and Human Services, and a few others which are not included here.

How do those magnitudes compare with what the nation as a whole is spending on R&D? One may look at trends based on information collected mostly by the National Science Foundation for the past decade. Table 4-1 shows funding levels in both current and 1981 constant dollars. The most striking characteristic of the tabulation is the rapid growth in industry-funded R&D, beginning in 1978. While in 1970, government funding was nearly 50 percent ahead of industry funding; by 1980, industry funding had caught up with gov-

TABLE 4-1. How Research and Development Are Funded: Sources in Fiscal Years 1970–81 (in billions of dollars)

Fiscal year	Federal government share[a]		Industry share		Universities colleges, other nonprofits		Total current $	Total constant $[b]
	($)	(%)	($)	(%)	($)	(%)		
1970	14.8	56.7	10.4	39.8	0.8	3.1	26.1	28.6
1971	14.9	55.8	10.8	40.4	0.9	3.4	26.7	27.8
1972	15.8	55.6	11.7	41.2	0.9	3.2	28.4	28.4
1973	16.3	53.1	13.3	43.3	1.0	3.3	30.7	29.0
1974	16.8	51.2	14.9	45.4	1.1	3.4	32.8	28.3
1975	18.1	51.4	15.8	44.9	1.3	3.7	35.2	27.7
1976	19.8	50.9	17.7	45.5	1.4	3.6	38.9	29.1
1977	21.7	50.6	19.7	45.9	1.6	3.5	42.9	30.3
1978	23.9	49.8	22.3	46.5	1.8	3.8	48.0	31.6
1979	26.6	49.1	25.6	47.2	2.0	3.7	54.2	33.3
1980	29.3	48.0	29.5	48.3	2.3	3.8	61.1	34.4
1981	32.7	47.3	33.9	49.1	2.5	3.6	69.1	35.7

Note: Percentages may not add to 100 due to rounding.

Source: Final Report of the Multiprogram Laboratory Panel, Energy Research Advisory Board, Support Studies, vol. 2 (Oak Ridge, Tenn., Oak Ridge National Laboratory, September 1982).

[a] Includes national laboratories.

[b] Current dollars converted to constant (1981) dollars through use of the GNP deflator series.

ernment; and, by 1981, it appears to have overtaken it. Unfortunately, without knowing the composition of industry funding it is impossible to judge what these trends imply. For example, much of industry R&D is aimed at product differentiation, designed to improve a firm's competitive position. It typically results more in wider consumer choice than in technological advances. This is often referred to as "defensive" R&D. Government funding is more likely to be aimed at fundamentals and true advances in science and technology, of the kind that by their nature profit-making entities shy away from.

The small share of academia in funding R&D is not surprising. Universities have been hard pressed for funds to carry on their education mission. Generating funds for R&D has been a problem. However, their share of the total has been roughly maintained.

A second way of looking at R&D is to determine not who *funds* but who *performs* R&D. The data provided in Table 4-2 tell a different story. Not only is the government the smallest performer of R&D, but its role has continuously diminished. The academic world emerges as the second largest performer, but still does less than one-fourth of what industry does. The largest part of government funding has consistently gone toward R&D carried on by industry—close to 70 percent in recent years. All told, the nation's R&D activities today amount to somewhat over 2 percent of the gross national product.

This, then, is the financial and institutional context in which energy R&D

TABLE 4-2. **Who Performs Research and Development:**
R&D Performance for Fiscal Years 1970-81 (in billions of current dollars)

Fiscal year	Federal government[a]		Industry		Universities, colleges, other nonprofits		Total
	($)	(%)	($)	(%)	($)	(%)	($)
1970	4.0	15.3	18.1	69.3	4.0	15.3	26.1
1971	4.2	15.7	18.3	68.5	4.1	16.5	26.7
1972	4.5	15.8	19.6	69.0	4.3	15.5	28.4
1973	4.7	15.4	21.2	69.3	4.7	15.4	30.7
1974	4.9	14.9	22.9	69.6	5.0	15.5	32.8
1975	5.3	15.1	24.2	68.8	5.7	16.2	35.2
1976	5.7	14.7	27.0	69.4	6.2	15.9	38.9
1977	6.1	14.2	29.9	69.8	6.9	16.0	42.9
1978	6.9	14.3	33.2	69.0	8.0	16.6	48.0
1979	7.5	13.8	37.6	69.4	9.1	16.8	54.2
1980	8.1	13.2	42.8	69.9	10.3	16.8	61.1
1981	9.0	13.0	49.2	71.1	11.0	15.9	69.1

Note: Percentages may not add to 100 due to rounding.

Source: Final Report of the Multiprogram Laboratory Panel, Energy Research Advisory Board, Support Studies, vol. 2 (Oak Ridge, Tenn., Oak Ridge National Laboratory, September 1982).

[a] Includes national laboratories.

is imbedded, an area that is characterized, as we shall see, by the existence of a large and unique governmental tool—the national laboratories.

ENERGY R&D

It would be most satisfactory if we could dispense with a discussion of overall R&D funding and performance and go straight to R&D in the energy sector. Unfortunately, the existing data are not satisfactory, in that government statistics include funds made available to industry, and (except for the total in 1979) the industry data do not show what portion of their effort was funded by the Department of Energy. Table 4-3 shows the breakdown of government R&D funding, and Table 4-4 shows industry R&D in the energy field.

Thus the two tables are not additive; nevertheless, they do give some insight into major allocations of R&D funds. One additional difficulty is worth mentioning as well: companies report what they consider to be R&D and, as a subcategory, what they consider to be energy R&D. Those familiar with the respective reporting industries surmise that much of what is reported as energy R&D is in reality engineering work. As a consequence, the numbers may be substantially inflated.

TABLE 4-3. Budgets for Energy R&D, for 1976, 1981, 1982
(in millions of dollars)

R&D category	Fiscal year 1976[a]	Fiscal year 1981	Fiscal year 1982
Defense	850	1,505	1,706
General science	246	504	543
Solar, geothermal electric energy, and storage	182	730	372
Magnetic fusion	167	383	454
Nuclear	817	1,165	1,140
Environment	205	227	173
Supporting research	145	285	293
Fossil	398	822	417
Conservation	39	207	81
TOTAL	3,049	5,828	5,179

Note: No data are available regarding energy R&D funding from other governmental sources. These budget figures are budget obligations in current dollars for the Department of Energy only.

Source: Final Report of the Multiprogram Laboratory Panel, Energy Research Advisory Board, Support Studies, vol. 2 (Oak Ridge, Tenn., Oak Ridge National Laboratory, September 1982).

[a] The earliest available year is 1976. The funding agency was then the Energy Research and Development Administration (ERDA), subsequently absorbed by the Department of Energy.

TABLE 4-4. Funds for Industrial Energy R&D Performance,
for 1973 and 1979 (in millions of dollars)

Category	Fiscal year 1973	Fiscal year 1979
Fossil fuels, total	438	1,312
Oil and gas	348	667
Coal	49	287
Shale	12	20
Synthetic	a	215
Other	29	122
Nuclear, total	501	917
Fission	476	798
Fusion	25	119
All other, total	70	1,667
Geothermal	1	160
Solar	2	347
Conservation and utilization	a	731
Other	67	429
TOTAL	$1,009	$3,896[b]

Source: Final Report of the Multiprogram Laboratory Panel, Energy
Research Advisory Board, Support Studies, vol. 2 (Oak Ridge, Tenn.,
Oak Ridge National Laboratory, September 1982).

[a] Not available.

[b] Of which $1,377 were from the Department of Energy.

A few characteristics of energy R&D stand out clearly. For one, until
the advent of nuclear energy R&D was carried on for the most part by energy-
producing and -consuming industries, with only a moderate amount of government activity. Thus the history of energy R&D was written not by government but by the great scientists and inventors. Many are remembered even in
the terminology of energy: Volta, Ampere, Ohm, Watt, Carnot, and others.
In the United States, to the extent that government carried on or even funded
energy R&D, it limited itself to broad traditional areas such as geology or
whatever was required to support its role in hydroelectric schemes situated on
federally owned river courses. There was research on miners' health and safety
in the coal mines, to be sure, but such topics as efficiency in combustion or
use, or availability of alternative sources were treated either by industry or in
learned institutions, or not at all. Largely because the relevant land was federally owned, shale oil research was one of the early R&D activities undertaken
within the government, but it constitutes the exception rather than the rule.
On the whole, it was the academic community that did the research and industry that provided the development, though with a growing participation in
research (of which the Bell Laboratories are probably the best example).

THE EMERGENCE OF GOVERNMENT AS FUNDER AND PERFORMER

The pattern of energy research has changed radically, however; first with the emergence of nuclear fission, and then with the 1973–74 oil shock. In the early days of the nuclear age government performed almost all existing R&D, spanning the entire range of science and technology, and only later was joined by others. The oil shock greatly broadened the scope of government R&D in achieving a variety of social objectives that lay only partly in the province of energy, such as the relationship of energy to foreign policy or to economic growth.

The establishment of several national laboratories during and following World War II marked the beginning of a new era in energy R&D. Here the government stimulated, funded, and organized an entire new branch of science that went all the way from gaining basic scientific knowledge to constructing a workable explosive device—the atom bomb. Then, looking beyond its destructive capacity, government researchers explored the atom's peacetime potential. Expensive facilities and the employment of top scientists and engineers were the hallmark of the national laboratories. Government research has never been the same, for the development opened wholly new vistas that were at once complex, expensive, and high-risk endeavors. Having taken the plunge, the government subsequently spawned laboratory after laboratory and program after program, all specializing in different nuclear topics and dispersed across the country. To be sure, the old army arsenals and navy labs played important roles, but their research continued to have few, if any, civilian applications.

For many years now, nine large multipurpose national laboratories have been in existence, employing, in the early 1980s, some 40,000 people with a combined annual budget in the neighborhood of $2.5 billion. Several smaller ones, each devoted exclusively to a single purpose or mission (such as coal, petroleum, solar), round out the governmental energy R&D program.

Facilities that were first devoted single-mindedly to the construction of nuclear weapons have in time been extended to include the many broader problems and issues, such as health and safety, that are linked to nuclear fission. This in turn, expanded the range of the laboratories' research into biology, chemistry, medicine, and other related fields.

Nor was that all. Two subsequent developments helped to expand the role of government and that of the national laboratories in energy R&D. One such development was the emergence, or rather, the recognition of the environmental problems associated with the production, transportation, conversion, and consumption of energy. Because these problems were "externalities"—not taken into account in market transactions by producers or consumers—the funding or undertaking of research became largely a responsibility of government, even though industry itself spent considerable effort on such R&D, as regulations required it to improve or control its polluting activities.

The second development—the 1973–74 oil embargo and the ensuing price revolution—dramatically underscored the need for alternative energy sources. Government involvement was required because it seemed that market forces alone could not be relied on to achieve rapid expansion of existing sources such as coal or to develop new ones such as solar energy. Government, accepting the challenge, both funded and carried out R&D in these areas. It did so quite naturally. For example, in devising criteria for environmental impact statements required for nuclear power plants, the Atomic Energy Commission early on had turned to the national laboratories for help. They could easily draw on their experience in biological and medical research associated with the health and safety aspects of nuclear power. And once such expertise existed, why not use it to assess the environmental aspects of coal-burning or other energy-associated processes? And, why not raise further questions—and try to find answers—as to how coal combustion could be technologically altered to pose fewer hazards to human health and land and water resources?

Other fields came in for federal funding, generally motivated by a need for government initiative in situations in which market forces were not likely to bring about adequate results. On the supply side, for example, a search was begun for a cleaner combustion process for burning coal, such as fluidized-bed combustion (in which coal is burned in a bed of ash and limestone that absorbs most of the environmentally noxious sulfur contained in the coal), the rationale being that private enterprise will be slow to devote itself to such research since the cost must be borne solely by the firm, while the benefit—cleaner air—will be a boon to society in general but will yield no direct return to the coal users. Therefore, industrial users would be unlikely to rush into the needed research and engineering work until they have been motivated by government regulations, and then it is unlikely that many will go beyond the point of meeting established standards.

More efficient energy use in buildings is another case in which few private entrepreneurs or organizations have found it in their interest to underwrite or perform this kind of research. The building industry is a badly fragmented one, having no history of research, and its few isolated efforts are poorly funded. The utility companies supplying energy to buildings are regulated monopolies, and as a rule they, too, do not engage in research. It is not clear that by doing research as individual firms they could recover their expenditures in the rate base set by public service commissions. If they were successful, their rates would presumably be lowered, and if they failed, they might be told that their R&D expenditures had been imprudent. More recently, the utilities have banded together to form the Electric Power Research Institute and the Gas Research Institute, and have sponsored wide-ranging research, largely aimed at improved technologies. Their funding is derived from special assessments on utility sales, sanctioned by the appropriate governmental bodies on either the federal or the state level, giving these bodies a quasi public character.

The danger exists that in the absence of a good way to compare benefits and costs, research will be funded or undertaken by the federal government, which in the case of energy means mostly the national laboratories, that could be done as well in the private sector. One research issue is linked to the next, and curiosity—as well as the natural desire to become independently expert rather than having to rely on the expertise of others—creates an environment for taking on a growing number of tasks. In response to some searching questions, it has become likely that some pruning—designed to cut back the government's performance of R&D to functions in which it possesses a clear comparative advantage—will take place. Such functions are connected either with the possession of unique and very costly facilities, built at government expense (often called a "national stewardship" or "trust" function); unique scientific and technological skills; the need for research that, to be credible, must not be tainted by the slightest perception of serving the industry's rather than society's interest; and the need for research in areas in which industry is uninterested or lacks the capacity for performing it. Much energy research fits the last description, thus justifying a unique role for government instead of private initiative.

GOVERNMENT R&D AND MARKET FORCES

While it is not too difficult to set up categories as those just listed, specific decisions as to where the comparative advantage lies are bound to be difficult and controversial. Some have offered a fairly simple rule-of-thumb: limit the government's R&D role to areas that have long gestation periods, carry a high risk (in the sense of uncertainty of success) and promise a high payoff if results are successful. This is a good start, provided it will admit of exceptions, especially in those fields that have poor prospects for finding alternative sponsors outside government. For a number of reasons discussed below, energy conservation research is a prominent example of such an exception.

Government-sponsored R&D in energy differs from two other areas in which government in the past has concentrated its R&D efforts: defense or weapons systems, and space exploration. In the latter instances, government not only provides the funds, it also writes the specifications and eventually owns and uses the resulting product or service. As a result, while costs play a part, overruns can be—and are routinely—accommodated by supplementary congressional appropriations. The taxpayer foots the bill, but no other damage is done. Both the space program and defense compete for appropriations, but neither competes in the marketplace for consumer acceptance and only in a minor way (for example, satellite communications) does either compete for international markets. The products of energy R&D, on the other hand, must prove their merit in a competitive setting; that is, they must be able to find their place in the market, even in cases where initial subsidies are used to introduce them. People continue to ask, If we can put a man on the moon, why can't we have a perfect substitute for Arab oil (or dirty coal, or risk-plagued nuclear power)? There are two answers. First, while we may succeed in having all of it *technologically,* that will do little good unless it is also economically efficient, environmentally benign, institutionally feasible, and politically acceptable. Second, the projects generally have had a single, clearly defined objective—to build a nuclear explosive device, to build a spacecraft, and so on. The objective of energy R&D is to help the country cope successfully with problems of energy supply and demand. That is a very different, multifaceted objective.

This qualification is crucial to understanding the controversy that surrounds the government's role in energy R&D. This controversy, as indicated earlier in this chapter, revolves around the appropriate role of government in the different phases of R&D, that stretch from basic research to commercialization. There is widespread agreement that the role of government is largest in research, and almost nonexistent in deployment. But there is much disagreement on just where government fades out as research results lead to development and demonstration. One basic consideration is that, since the resulting product—and most energy R&D is aimed at hardware—must be competitive in supplying energy for private use, the private sector should be involved early on. In this respect, it is also important to curb the government's tendency for zeroing-in prematurely on a specific technology or design (as it did, for example, in the case of the nuclear reactor that owed its present form largely to the success of the reactor developed for nuclear submarines). In the early stages of R&D *diversity* is most important. While it is clear that deployment must be undertaken by the private sector, the decision is less obvious at the preceding—demonstration or pilot plant—stage. Here, again, a major role should be played by the private sector, but companies may be unwilling to be involved without government participation or safeguards against excessive financial hazards. Government, on the other hand, once it is financially involved, often tends to take a hand in the execution of the project. This was true, for example, in the synfuels projects sponsored before creation of the Synthetic Fuels Corporation.

It is perhaps simplistic to expect hard and fast rules, by which everyone will play, in all types of R&D. In the early years of the post-1973–74 period government tended to operate very far forward in the R&D chain. It not only funded, but actively participated, in the planning and execution of projects that eventually would result in new industrial ventures. More recently, the government has been withdrawing from such enterprises, except, of course, where industry continues to be uninterested and incapable (for example, advanced nuclear technology, such as the breeder or fusion). In general, the Department of Energy has pulled out of anything having the characteristics of a pilot or demonstration plant. This has been especially noteworthy in regard to synthetic fuels. The Synthetic Fuels Corporation (SFC) has been created by Congress to help speed up the timetable of this particular technology. In that endeavor, however, the intermediate learning stage, that is, constructing facilities short of full-blown production plants, could become an orphan. It might be no longer of interest to government and too risky an enterprise for the private sector. The SFC assists by making or guaranteeing loans, by purchase guarantees at stipulated prices, and in other ways. But while it does select from prospects proposed by the private sector, and can negotiate details, the SFC does not involve itself in the actual R&D. Moreover, the lull in oil price rises that began in the early 1980s has taken much of the steam out of this particular venture, as the private sector reviews its perception of the future market for synthetic oil and gas and takes a less venturesome stance. All this could change quickly, with yet another turn in the outlook for oil, but as of late 1982 this particular segment of R&D highlights the problem of heavy reliance on the private sector.

CONTENT AND DIRECTION OF ENERGY R&D

Among the many areas in which energy R&D is being conducted in the United States and abroad, we have selected only a small number that seem especially important.

Nuclear Energy

The pace of progress of nuclear power generation as part of the nation's energy supply has fallen far below earlier expectations and is unlikely to pick up speed soon. Nuclear R&D nevertheless has been proceeding along several lines, most of them both funded and conducted by the federal government. One segment is concerned with improving the nuclear power-generating system as it now operates. It involves principally improved safety, greater efficiency in the utilization of uranium, less costly enrichment of uranium in the process of fuel fabrication, and safer methods of disposing of radioactive waste. In the wake of the 1979 Three Mile Island accident, research on more efficient and reliable instrumentation, as well as on personnel training and qualifications, has been emphasized. As existing plants age (one sometimes forgets that the oldest nuclear power plant was built in the early fifties), disposal of entire plants also is becoming an R&D topic, as are materials problems (such as embrittlement) that show up after prolonged exposure of equipment to a radioactive environment.

A related effort concerns other types of converter reactors, but U.S. interest in these is only modest. Neither gas-cooled reactors nor those like Canada's, that utilize natural uranium, have much of a following. Uranium scarcity has not been an issue for some time and, given the slow growth of the industry and increases in uranium reserves, is unlikely to become one soon. Thus improved efficiency and, to a lesser degree, more uranium-saving enrichment take a backseat to greater safety measures and to disposal of nuclear waste. The latter problem urgently needs to be resolved. The absence of an acceptable disposal method that has, in fact, been put into practice is bound to impede expansion of nuclear power. It has been of concern in several states, and progress has been slow despite substantial evidence that disposal technologies are available, that the chances of escape of radioactive substances from deposits are small, and that if escapes should take place, they are unlikely to have a severe impact. Thus, the major barriers to resolving the problem are bureaucratic or institutional rather than technical: poor and indecisive management and a lack of consensus on location of disposal sites and methods. Legislation passed in late 1982 may be an important step in creating an orderly procedure, though states will retain a strong voice in the crucial matter of location decisions. Nonetheless, one must caution that finding a geologically satisfactory site has been turning out much more difficult in practice than theory would have led one to believe.

Closely related is research on the reprocessing of spent fuel. This phase of the fuel cycle has been highly controversial since it involves the separation of plutónium—the stuff from which explosive devices can be made—from the fuel. Several other countries have moved ahead of the United States in this field and have begun reprocessing on a commercial scale. Whether spent fuel will eventually be disposed of as such or only after it has been reprocessed to separate out those components that can again serve as fuel remains to be settled, and it is up to ongoing research to contribute answers.

Uranium enrichment, that is, raising the amount of the fissionable uranium 235 in uranium, can be accomplished in several ways. R&D efforts are aimed primarily at doing it with a smaller investment of energy and, at the same time, possibly with less cost. The principal new technology available uses lasers to separate the uranium-235 isotope from the other components of the uranium atom. A major concern is that the method might eventually be made so simple that it will provide wide access to the technology and thus to the uncontrolled fabrication of explosive devices. Considerable government funding continues to be channeled into this approach.

A different line of R&D is aimed at nuclear reactor *concepts* not yet in commercial use or even demonstrated as feasible. Of these, the breeder and fusion are the most important, and of the two, the breeder is, and has always been, the more controversial. The main argument against the breeder has been that it is fueled by plutonium, the weapons material par excellence, and that possible lack of competitiveness and safety make it even less advisable to incur the risk of spreading this material around. At the other end of the spectrum are those who see in the breeder the principal justification for pursuing nuclear power generation at all. Only the breeder makes full use of the enormous energy potential per unit of uranium and holds out the prospect of a virtually limitless supply of energy. Without the eventual emergence of the breeder, the current generation of fission reactors might be considered a mere interlude, lasting a few more decades, in the long span of history.

The stakes in breeder R&D are thus high. A successful, competitive breeder reactor would go a long way toward providing energy for the world, perhaps for centuries. On the other hand, widespread political instability, both internal and international, gives importance to withholding additional and disastrously destructive tools of war and terror from those anxious to use them. Here the battle is joined. R&D has two aims: to permit the building of a power-generating reactor with costs in the competitive range; and to perfect technological and institutional measures that will minimize the opportunities for diversion of plutonium. It is worth mentioning here that the decisions are not all in the hands of the United States. For example, France is far ahead of the United States in perfecting a breeder, and so reportedly is the Soviet Union. If the French endeavor is successful, there will be an interesting debate in the United States: the choices will be to license and employ the device and "ride piggyback," so to speak, on a foreign engineering accomplishment, to pursue indigenous R&D, or to turn one's back on both. French efforts to market the breeder abroad would add yet another issue.

One thing appears reasonably certain. The need to develop the breeder quickly has receded, because of the greatly reduced rate of growth in energy consumption in general, electricity consumption in particular, and the much longer time horizon for uranium supplies caused by the slowdown in nuclear power growth. For much of the 1970s, there was concern that the rapid buildup of the nuclear power industry would strain uranium supply, quickly drive up its price, and make the breeder both needed and competitive given its highly efficient uranium use. Today these concerns have abated, but the concept of the breeder is too intriguing for the technology to be discarded. Breeder R&D will no doubt continue but probably at lower speed.

Paralleling the lines of R&D discussed, there is a field of research that focuses on safeguards, that is, on the ways and means of minimizing the chance that nuclear material may be diverted from the fuel cycle and employed in weapons-making. This is heavily a worldwide rather than a U.S. problem. Progress would ease the continuing dilemma faced by the nuclear weapons states of having to discriminate against those countries who wish to obtain nuclear reactors but who are deemed to be potential weapons makers (see further discussion of this in Chapter 7).

Fusion power has long been looked at as a kind of "ultimate energy source," largely because it can be fueled by components (isotopes) of hydrogen, and thus can be said to have a virtually inexhaustible fuel supply, although the need for some other materials, such as lithium or beryllium, may set limits. From what we now know fusion is also likely to have fewer environmental problems, and its complexity reduces the potential for clandestine weapons manufacture.

How Reactors Are Fueled

The basic fuel for the currently employed electricity-generating technologies based on nuclear fission is uranium. In the type of reactor produced by U.S. manufacturers and employed by U.S. electric utilities, the component (isotope) of uranium that sustains the fission reactor is uranium 235. Unfortunately, only 0.7 percent of a unit of uranium consists of this component. For the fission reaction to occur and be sustained in the current type of reactor, a material that contains at least 3 percent uranium 235 is needed. A process called enrichment performs the function of increasing the uranium-235 content up to the required minimum.

What is left behind as not useful in presently commercial types of reactors is the other uranium component (isotope), uranium 238. It is stored to be available in the breeder, should that technology mature to commercial scale. The breeder can be fueled by a variety of combinations of uranium and plutonium, and also by thorium. Its main objective is, concurrently with producing heat for electricity generation, to convert material that will not fission (like uranium 238) into material that will fission (like plutonium). When in the process of being used up, the original fissionable materials in the reactor create a larger number of newly fissionable atoms, then we have a *breeding* process. Of course, even that operation eventually comes to a halt, but not until the nonfissionable uranium 238 (which is useless in the conventional reactor) has been used up (see Figures 4-1 and 4-2). Other types of reactors are feasible. One of them uses the metal thorium rather than uranium. Others use natural, that is, not enriched uranium. Then there are more efficient reactors, called advanced converters that exploit more of the uranium fuel but none come near equaling the breeder.

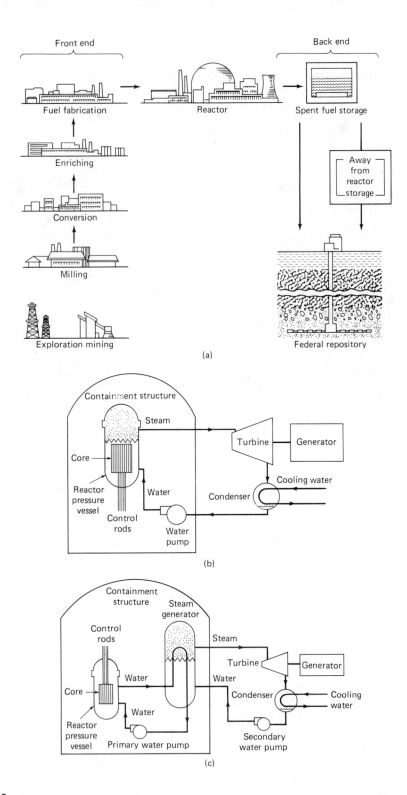

Front end

Back end

Fuel fabrication

Reactor

Spent fuel storage

Enriching

Conversion

Away
from
reactor
storage

Milling

Exploration mining

Federal repository

(a)

Containment structure

Steam

Turbine

Generator

Core

Cooling water

Reactor
pressure
vessel

Water

Condenser

Control
rods

Water
pump

(b)

Containment
structure

Steam
generator

Control
rods

Steam

Turbine

Generator

Core

Water

Water

Condenser

Cooling
water

Reactor
pressure
vessel

Water

Primary water pump

Secondary
water pump

(c)

As the name implies, fusion generates energy from the fusing of atomic particles rather than from their fission. The practical difficulties of achieving fusion for peaceful purposes are enormous. They have to do largely with the difficulty of confining the enormous heat and power that is generated. Also, a great deal of energy is needed to generate the fusion reaction. That, plus losses during the process, detracts somewhat from the attractiveness of this technology.

Spending for nuclear fusion R&D, for which the government is practically the sole funder, recently has been in the range of $500 million a year or ten times the amount spent in the early 1970s. It seems assured of ample long-range funding, although, unlike the breeder, even the concept has not been proved. That is to say, as yet no energy has been generated on the laboratory scale. Only pieces of the process have been successfully demonstrated in government-funded or -owned laboratories that are as costly as they are complex. It is hard to define the current stage of progress. An often-heard view is that within ten years or so we should know whether fusion will ever be a workable power-generating technology. On either side of this view are believers and skeptics. There is also the question of whether the country can afford, and if it is wise, to pursue both the breeder and fusion, since both hold out a similar promise—a practically unlimited supply of electric energy, or of energy generally if one anticipates an "all-electric" future. Those arguing against fusion stress that the breeder is achievable, whereas fusion might not be. At the same time, to cease fusion R&D might deprive us of an abundant source that has safety advantages over the breeder. The only noncontroversial aspect is that in both instances R&D is a task in which government must play the leading role. Risk, costs, and the overall time frame make it wholly unattractive for the private sector to carry much of the burden.

Progress made in nuclear R&D will have a heavy bearing on the competitive struggle between nuclear and coal-based power generation. While nuclear reactors built some time in the past appear to produce electricity at substantially lower rates than do coal-fired ones, newly built reactors may not have much of an advantage, if only because of the high cost of capital that punishes capital-intensive facilities (see box, page 116) and the failure to shorten the period between the decision to build a plant and the time it begins operations. Obviously, the longer funds are tied up prior to operation, the higher the eventual cost of the plant.

Figure 4-1 Elements of the light-water reactor. (a) Light-water reactor fuel cycle; (b) schematic diagram of a boiling-water reactor power system; (c) schematic diagram of a pressurized-water reactor power system. Sources: (a) From the League of Women Voters Education Fund, *A Nuclear Power Primer: Issues for Citizens* (Washington, D.C., October 1982); (b) and (c) From *Nuclear Power: Issues and Choices,* © 1977, The Ford Foundation. Reprinted with permission from Ballinger Publishing Company.

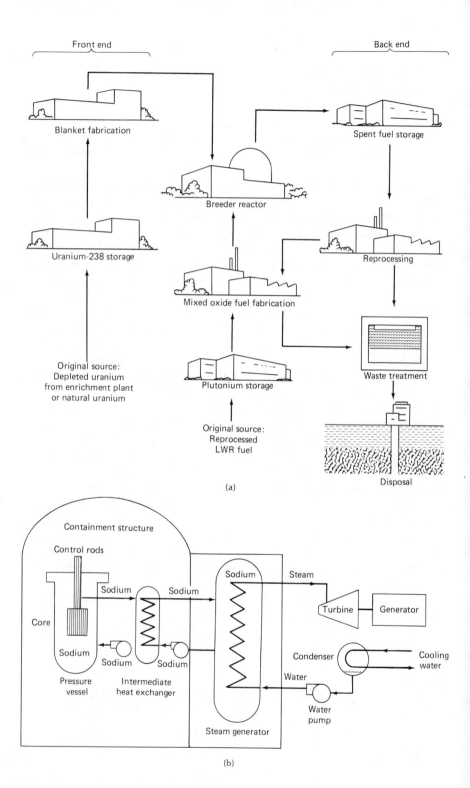

(a)

(b)

Synthetic Liquids

Given the continuing need for a large liquid fuel supply, especially to meet the needs of the transportation sector, and because oil security represents the most vulnerable aspect of the U.S. energy situation (and of most other countries as well), the search for an oil substitute has been a perennial R&D topic. Principally research has stressed the conversion of coal and oil shale into a liquid fuel that could meet the needs now being served by petroleum products. The means to accomplish this conversion are not unknown. In various parts of the world, and at various times in the past, conversion plants have been in operation. Some, like the SASOL plant in South Africa, have been turning coal into liquid fuel for some time. Two tar sand plants have been operating in Canada, one for more than a decade, and others are planned to come on-line. More than one hundred years ago, shale oil operations were carried on in Scotland, and later in Sweden as well.

Then what is the problem? One is scale. Neither the South African nor the German synfuels plants in operation during World War II—often cited by enthusiasts as evidence of their practicability—begin to compare in magnitude with those needed if synthetic liquids are to make a significant contribution in the United States today; and one should note that greater scale brings greater problems.

Second, none of the earlier plants were built during an era of environmental awareness. The required conversion processes are likely to generate substances that are adverse to human health. Like many other such hazards, they are probably manageable, but at a cost and only after time-consuming research has been undertaken.

Third, operating experience is limited to one or two processes. A number of others appear attractive because they are less costly, produce a preferable mix of products, seem less dangerous, and are affected with fewer environmental problems.

Fourth, these will be very large plants, and as such they will place additional stress on local resources, both natural and human, and involve especially complex issues in regard to shale. Thus, one may expect the site-selection process to be both long and agonizing. For these and other reasons, the cost of the proposed facilities and of the products that they produce must remain highly uncertain until the first full-sized plant has been in operation for two or three years. Even in the planning phase estimated costs have undergone

---

Figure 4-2 Elements of the breeder reactor. (a) Breeder reactor fuel cycle; (b) Schematic diagram of a liquid metal fast breeder reactor power system. Sources: (a) From the League of Women Voters Education Fund, *A Nuclear Power Primer: Issues for Citizens* (Washington, D.C., October 1982). (b) From *Nuclear Power: Issues and Choices,* © 1977, The Ford Foundation. Reprinted with permission from Ballinger Publishing Company.

The Significance of Capital Intensity

Some energy technologies require a very heavy input of capital to begin with but function with only a small amount of labor and other operating expenses. Nuclear power stations, natural gas pipelines, synthetic fuel plants (if and when they will come into existence) are examples discussed in various contexts in this book. Capital intensity has a number of important consequences:

- The rate of operation can spell profit or loss. Since paying off the capital is the major expense that must be met no matter what, operating at a low rate of capacity will depress net income, operating near capacity will boost it. This explains, for example, why gas pipeline owners favor low national gas prices: they would promote greater throughput.

- In times of high interest rates, often put as "high cost of money," capital-intensive ventures are at a disadvantage. Thus nuclear power plants, in which capital expenditures are large, are looked at askance in comparison with coal-, gas-, or oil-fired plants that have a lower capital input.

- When output, or throughput, declines, a capital-intensive operation faces the choice of incurring a financial loss or raising the price at which it sells its goods or services. As electric power consumption declined, customers, who thought they had done the right thing by curtailing consumption, were outraged when they faced higher bills for less electricity. Yet, that is the way in which capital-intensive operations function: a fixed cost is spread over a larger or smaller number of units sold.

- Capital-intensive technologies are on the whole likely candidates for government assistance to provide, or at least guarantee, the heavy initial outlay. This not only opens up complex issues of how the private sector interacts with government, but also causes long delays both on substantive and procedural grounds. As the time during which funds are tied up and need to yield a return lengthens, capital costs increase further. High capital costs thus tend to beget even higher capital costs. The widespread notion that big projects tend to bog down because of government interference and bungling needs a little correction. They tend to bog down precisely because the initially heavy capital investment virtually makes government involvement inescapable. All else follows.

continuous escalation, and any cessation of oil price increases would greatly dampen the synfuels fervor. Meanwhile, the inherent problems of synfuels manufacture require continuing R&D.

An intermediate product for some of the intended plants will be gas, a product more easily derived than liquid. Stopping with gas as the end product would simplify the process, but the relief from oil depletion would then have to come indirectly, that is, through the substitution of synthetic gas in uses now met by oil. This would help; but it would cause substantial complications for the mix of products supplied by oil refineries and, on the whole, would retard the pace of reduction in oil use. No doubt, synthetic liquids are more desirable.

The history of shale oil well illustrates the vagaries of specific types of R&D. As explained in Chapter 3, there is no mystery about the fact that oil can be extracted, or—to use a more apt term—*distilled* from oil shale. As long as sixty years ago there was speculation bordering on confidence that as petroleum was depleted, the United States would turn to shale oil. But conventional petroleum reserves instead experienced vast and unceasing expansion, first at home and then abroad, especially in the Middle East. Since petroleum was cheap, further attempts to study the characteristics of oil-bearing shale and the least costly way of extracting its oil turned erratic.

During the 1970s the need for a petroleum substitute became more urgent. It then was discovered that the costs would be much higher than earlier had been anticipated. Moreover, by that time new obstacles had arisen—above all, opposition to disturbing the natural environment in which oil shale is located. Concern focused on damages to clean air and water from the voluminous wastes left behind, and on the consequences of introducing a sizable labor force into parts of the country that previously had a low population density. Currently, the needed investment ranges from $3 to $5 billion or more for a daily plant capacity of 50,000 barrels of oil. Even at the lower estimate it takes giant corporations or government, or a combination of both, to command and risk that kind of capital. When one considers the uncertainties of a new technology to be used on an unprecedentedly large scale, the environmental problems, and the uncertain competitive position at the time the facilities are completed, it is not surprising that shale oil R&D has had a roller-coaster ride. With the easing of oil prices in the early 1980s, interest in pursuing both coal and shale conversion was again on the wane.

Coal Combustion

Abundant coal is looked to not only as an eventual source of liquid fuel, but also as a fuel that, if burned with an environmentally less noxious impact, can replace oil directly in diverse uses. R&D designed to improve the combustion process has been carried on for many years. The main focus has been on capturing the various polluting elements in the stack gas, but especially "soot," or particulate matter, and the sulfur contained in the coal. Current technology for sulfur control through so-called scrubbers—large, somewhat temperamental, and costly pieces of equipment that strip the stack gas of most of its sulfur dioxide—has been available for some time. The Clean Air Act Amendments of 1977 made their use mandatory for all new coal-burning power plants. Scrubbers have three major drawbacks. Their performance has been spotty; they are costly additions to the plant, further raising the price of electric energy, which for all practical purposes precludes their use for smaller coal-burning facilities; and they leave behind a residual waste that is hard to manage and dispose of.

Burning coal on a bed of lime and ash and with a continuing blast of air to agitate the mix, a method known as fluidized-bed combustion, overcomes most of these difficulties. The residual waste is easily handled, and the process seems sturdy. However, upscaling to larger units is seen as a problem, costs are still high, and currently we have no operating experience in standard-sized coal-burning power plants of, say, 200- to 500-MW capacity. Government's R&D role has been an active and apparently effective one. As of 1982, nearly one dozen companies were marketing small-capacity boilers to users other than electric utilities.

Solar Energy

Relatively simple in its technological aspects, the central problems of solar technology are efficiency of conversion, cost of materials, and ways of introducing it into the current mix of energy sources. Solar energy offers both short- and long-term opportunities and can be exploited by individual users, commercial enterprises, and by government. As discussed in Chapter 3, the subject is beset by heavy emotional overtones and more than other sources lends itself to being misunderstood. For instance, it is widely assumed that, as sunshine is free, solar energy must be inexpensive. What is ignored is that it takes materials to capture the sun's energy and that the conversion process is very inefficient. Figure 4-3 shows how little of the sun's potential in fact is captured in the case of photovoltaics. Similar inefficiencies obtain in other uses.

For the relatively short period that the nation has paid serious attention to solar energy (and we use the term here in the narrow sense, meaning the direct use of solar radiation), several lines of R&D have evolved. One, *passive solar energy,* merely aims at getting the most out of solar radiation by capturing it through architectural design, location, position and control of windows, insulation, and so on. A second, *active solar energy,* uses special equipment, such as collector plates, on rooftops or on the ground, tubing to carry sun-heated water, pumps, or fans to distribute it, and instrumentation. A third approach is characterized by *photovoltaics,* which utilize characteristics of certain materials to cause solar rays to produce an electric current.

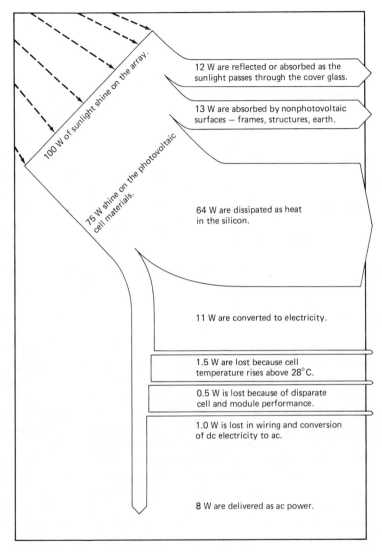

12 W are reflected or absorbed as the sunlight passes through the cover glass.

13 W are absorbed by nonphotovoltaic surfaces — frames, structures, earth.

64 W are dissipated as heat in the silicon.

11 W are converted to electricity.

1.5 W are lost because cell temperature rises above 28°C.

0.5 W is lost because of disparate cell and module performance.

1.0 W is lost in wiring and conversion of dc electricity to ac.

8 W are delivered as ac power.

100 W of sunlight shine on the array.

75 W shine on the photovoltaic cell materials.

Figure 4-3 Losses in photovoltaic energy conversion. Losses in six main categories reduce the overall efficiency of photovoltaic energy conversion. The figures shown here are representative of today's nonconcentrating silicon system performance. In general, the largest loss categories offer the most room for improvement, so they are the major topics of R & D. Cell loss is deceptive, however. For the 75 watt entering the silicon, the 64-watt loss (and 11-watt output) results from a 15 percent conversion efficiency. Even silicon's maximum theoretical efficiency of 22 percent would entail a loss of 58 watt. To do better requires photovoltaic materials and configurations that react more effectively to a greater range of the solar energy wavelengths. Source: *EPRI Journal* (December 1981), Electrical Power Research Institute, Palo Alto, California.

And there are others still—for example, *solar ponds,* in which especially engineered ponds are used to store heat over extended periods.

Of these, the first two are making steady progress, though the slow pace of new housing starts, and various difficulties in retrofitting the existing housing stock—that is, adding equipment or rearranging it after construction—slow down the speed of progress. Moreover, in many parts of the United States, solar energy can provide only part of the required heat so that a supplementary energy source or storage must be available to provide twenty-four-hour service all year round. Under those circumstances, the costly up-front investment in solar is made with less enthusiasm. Indeed, some believe that the house-by-house approach may turn out to have been a mistake. Clusters of houses might have access to solar techniques in less costly ways—for instance, through the above-mentioned solar ponds—though this in turn would impose an important locational constraint. R&D in both passive and active solar energy is focused less at the research end and more at the development, demonstration and deployment end. What works, at what cost, under what conditions, and with what kind and degree of maintenance—these are the more relevant questions. So is the matter of finding reliable advisers and equipment suppliers. Where R&D is least effective is in determining what does or does not motivate potential consumers from making the transition to solar energy use.

The field of photovoltaics, on the other hand, has greatly benefited from research. Industry (including some very large companies) has taken an interest and developed it rapidly. The principal task here is one for which industry is well equipped, namely, to discover and engineer low-cost materials, both for the solar cells themselves and for the supporting structure that holds the cell arrays, in order to offset the low energy yields that characterize this conversion process (see Figure 4-3). Important advances have been made in the manufacturing process in the last few years. Added to the conventional batch processes, in which individually produced crystals were sliced to provide the cell material, is a continuing process in which a silicon film deposited is produced capable of exciting a current when hit by a source of light. These advances no doubt have benefited from experiences gained in related high-technology fields such as semiconductors, transistors, integrated circuits, and the like.

Some believe that industry will continue to reduce the cost at which the arrays supporting the cells can produce electric power competitively, and that the time horizon for this is ten years rather than twenty-five. Others are more skeptical, citing the discouraging outlook for discovering a radically cheaper support material. All told, it is a technology that has been shown to work. The outlook is one of letting R&D bring down the cost bit by bit, as has happened quite successfully in the past. If we are lucky, cost reduction may come in one or two big jumps, based on new concepts or new approaches. But one cannot bank on such developments. Usually, initial cost reductions come quickly and easily; the final ones are the tough ones.

The manner in which the projected technology will be used, and, therefore, the extent to which it will relieve pressure on conventional sources of energy is not entirely clear. Very large arrays of cells are needed to produce modest amounts, measured in terms of conventional steam power plants. Space configurations must be evolved, but use in small, self-contained units seems one likely application. Another would be the coupling with conventional power plants to provide an intermittent alternative source. When used in isolation, backup power or storage is required; when the sun does not shine—at night, on dark days, and in winter—a supplementary source is required. Research on storage is thus an integral part of this as of all other systems that depend on solar radiation and operate independently of utility systems. Here too, both government and industry have carried on R&D.

Mostly, government-funded R&D, in cooperation with electric utilities, is going forward on other ways of generating electricity from solar rays. Bowls, troughs, mirrors are combined in various ways to concentrate sunlight on fluids in order to produce temperatures that permit generation of steam and electricity. The sums allocated are small, indeed miniscule, when compared with those devoted to fission and fusion, but to a large extent that is the nature of the beast. These machines or installations are not especially costly, and what is needed most is experimentation and gathering of operational results, one at a time. Once proved efficient and competitive, they could be rapidly multiplied and installed. In that sense, the next few years could be critical.

RECENT TRENDS IN GOVERNMENT-FUNDED ENERGY R&D

In a fluid area like R&D it is difficult to generalize about the size and characteristics of the effort. Both are time-specific. In 1982, to illustrate, with budgetary constraints and strong emphasis on the opportunities offered by market forces, the government's energy R&D budget underwent close scrutiny and substantial change. Its main thrust was a growing shift toward the electricity option, especially the nuclear variety, while R&D for all other energy forms as well as conservation was being scaled back drastically or phased out altogether.

According to the information brought together by the Energy Research Advisory Board of the Department of Energy (November 1981) of federally funded energy R&D (other than that devoted to basic science and technology), electricity took 63 percent in FY81 and 77 percent in FY82. And, within this category, nuclear rose from 50 to 72 percent between the two years, confirming the concentration on fission and fusion research. By contrast, R&D on solar dropped from 18 to 8 percent, and research on utilization and conservation from 20 to 11 percent. The trend is likely to further entrench the primacy of electric power research, and of nuclear within it, while solar, geothermal, fossil fuels, and conservation R&D are likely to lose ground, at least for a while. Considering the magnitude of the problems and the cost of operating anything at all, the resulting appropriations for these segments will be barely enough to keep these activities alive as federal government tasks.

Such fluctuations are to be expected when a government's philosophy changes. In this instance, the reductions were justified on the basis that the private sector stands ready to take on the activities given up by government, and that economic recovery will do the rest. This is a matter of substantial controversy. It is difficult to see, for example, how those programs that can make a contribution to greater efficiency in energy use will be picked up by private interests, even under the stimulus of economic recovery.

The most telling example is R&D in efficiency of energy use in the construction industry, especially in residential building. This was touched on earlier in the chapter. Several factors are at work to block R&D in energy conservation, but the most important is the multiplicity of actors—land developers, builders, architects, contractors, suppliers of appliances and hardware, real estate firms, owners, renters, landlords, municipal regulators—each of which has a different motivation. The one thing they share is that they are not set up to undertake research, and sellers are often not confronted with eventual users, so that the demand for efficiency has difficulties in becoming effective.

This is one of the areas where the market is not a force for R&D, or where the primary motivation for R&D is not to ease the energy situation, but to increase sales and profits. Easing of the energy situation would, of course, mask any setback that one might otherwise ascribe to a reduction in government funding of R&D. Moreover, much of R&D in which government is retaining a high interest has very long time horizons, so that one will not be in a position to assess its payoff, or even the wisdom of undertaking it, for quite some time. It is not only conceivable, but likely, that successive administrations will entertain different philosophies and will alter the thrust and mix of energy R&D. It is unlikely, however, that they will have more than marginal impact on the few massive R&D programs in which government sets the tone. They would seem more or less immune to the vagaries of the budgeting process, at least for the next few years. International competition alone would seem to favor their continuation. The troublesome areas are the smaller, less glamorous programs whose termination may kill off a promising solution to a pressing problem. These may each be small increments to coping with a large problem; but in the aggregate the contribution of a sizable number of small improvements—the better mouse trap—is not something to be dismissed lightly when compared with the more grandiose visions of a carefree energy future carried on the wings of the multibillion-dollar ventures.

BIBLIOGRAPHICAL NOTE

The literature on the technical aspects of energy production and use is vast, and much of it would qualify as relevant to research and development. However, relatively little is written on the nature, magnitude, economic, and institutional contexts of R&D as such, and of its energy component.

Those interested in getting a grasp on the overall organizational and institutional aspects of energy R&D will find the relevant chapter in *Energy: The Next Twenty Years,* a study sponsored by the Ford Foundation and administered by Resources for the Future (Cambridge, Mass., Ballinger, 1979), possibly hard going but rewarding. No less demanding are the reports issued at irregular intervals by the Energy Research Advisory Board of the United States Department of Energy. Of special interest is a report to the Secretary of Energy entitled, *Federal Energy R&D Priorities* (Washington, D.C., U.S. Department of Energy, November 1981). On a broader scale, the International Energy Agency in Paris has been releasing a report every two years in which it assesses the progress made in the member countries with regard to energy R&D. The latest issue is entitled, *Energy Research, Development and Demonstration in the IEA Countries.*

Finally, attention must be called to two relatively new organizations that were set up specifically to perform and promote R&D in the energy field. These are the Electric Power Research Institute (EPRI), located in Palo Alto, California, and the Gas Research Institute (GRI), located in Chicago, Illinois. As regulated industries, firms in these segments had little incentive and no institutional basis for undertaking R&D. For the opposite reason, there are no analogous organizations in either oil or coal. Both EPRI and GRI issue a journal and a large number of research reports in their respective fields. Lists of their publications are directly available from the two organizations.

Because it is still in a relatively early phase of development, nuclear energy has spawned a large literature in the R&D area, most of it aimed at the specialist. For the general reader, several of the books mentioned in connection with other topics, that is, Hans H. Landsberg and others, *Energy: The Next Twenty Years,* a study sponsored by the Ford Foundation and Administered by Resources for the Future (Cambridge, Mass., 1979); *Energy in America's Future: The Choices Before Us* (Baltimore, Md., Johns Hopkins University Press for Resources for the Future, 1979); *Energy Future* (New York, Ballentine, 1980); *Nuclear Power: Issues and Choices* (Cambridge, Mass., Ballinger, 1977); and the National Academy of Science's CONAES report—*Energy in Transition, 1985-2010,* Final Report of the Committee on Nuclear and Alternative Energy Systems (San Francisco, Calif., W. H. Freeman, 1980), still represent the most accessible sources of additional reading.

CHAPTER **5**

Competition
and Regulation
in Energy Markets

The enormous size of American oil companies makes them highly conspicuous. Six of the nation's ten largest corporations (measured in dollar sales) are energy firms. Exxon, whose sales totaled $108 billion in 1981, is nearly twice the size of General Motors, the biggest firm outside the energy field. Eight of the ten most profitable U.S. firms in the same year were energy companies, though they all ranked behind the American Telephone and Telegraph Company, which earned nearly $7 billion. No wonder that during the 1970s when energy prices were rising rapidly, consumers, businessmen, and politicians looked with suspicion on the giant U.S. multinational oil firms. Their size, their profits, their dealings with the OPEC countries, and just their omnipresence called their operations into question.

But on closer look, one sees that what is conspicuous is not necessarily what is most important. In a world of giants, size may confer little advantage; Mobil and Gulf are not intimidated by Exxon. A profitable past is no assurance of a profitable future, as the oil firms were reminded in 1982 when a recession and a supply glut drove profits down. Oil, though indispensable, is not the only source of energy, and medium and small firms flourish in other parts of the industry that have not attracted the giants. Size, in short, attracts attention but it is only one factor that affects how energy markets work.

These markets, as shown below, are remarkably diverse—displaying contrasting patterns of competition and monopoly, subject to varying degrees of government regulation, and ranging geographically from urban neighborhoods to the entire world.

The contrasts are familiar and striking. The retail market for gasoline is competitive, sometimes fiercely so; if gasoline is plentiful and business is slack, service stations cut prices or offer bonuses to stimulate sales. But there is little competition in the market for electricity; a single power company generally supplies the needs of an entire community, and sharp changes in economic conditions may have little effect on prices paid by consumers.

Coal mining is a relatively competitive activity; historically, although the industry has its share of large firms, small producers have moved in and out of business depending on changes in demand. Distribution of coal to market, however, is a different story; producers have been primarily dependent on railroads, whose schedules and rates have been tightly regulated by the federal government.

Geographically, the market for crude oil is worldwide and dominated by giant companies; for firewood it is local and unorganized, often with room for haggling between individual consumers and itinerant vendors.

Firms within energy markets are similarly diverse in their structure. Some companies operate in a single market—coal mining, for example. Others, notably the major oil companies, perform a sequence of functions that place them simultaneously in several markets. They search for oil, produce it, transport it, refine it, and distribute final products to consumers and businesses. In the economist's lingo, they are "vertically integrated" from wellhead to pump.

Some firms are "horizontal combinations," producing products that compete with one another. They may produce oil and coal, for example—two fuels that for some purposes are direct substitutes for each other.

Some firms are "conglomerates," combining not only characteristics of vertical and horizontal integration, but also including corporate divisions devoted to operations in other industries. As shown in Chapter 1, more than 17 percent of the assets of major energy firms are invested in other businesses—some of them in related fields, such as chemicals, and others in unrelated industries, such as retailing or manufacturing.

Why do different types of energy markets and firms exist? What kinds of changes do they face? What role should the government play in strengthening energy markets or in providing alternatives to them?

THE MARKET: CONCEPT AND ROLE

Although it is common to talk of a market as a *place* where goods and services are bought and sold—securities on Wall Street and meat in Chicago or Omaha—it is unduly restrictive to think of a market simply as a place. The world oil market is not established in a few locations, such as Saudi Arabia or Rotterdam. It encompasses transactions negotiated throughout the commercial world. The *market* is a term, in short, that signifies all the various kinds of arrangements that have been devised for bringing together buyers and sellers. These arrangements determine why sectors of the business system work as they do—with greater efficiency in some situations than in others, with greater fairness in some circumstances than in others.

Market boundaries depend not only on their geographic dimension but also on the nature of the product. Economists conventionally define markets in terms of "product substitutability." Two or more products may be considered to be in the same market if they can be substituted for one another. Thus an electrical utility firm may have the choice of building a plant that can be fueled by coal, oil, gas, nuclear fuel, or hydropower. In this case, all of these sources of energy may be viewed as being in the same market. Once a nuclear plant is built, however, the utility is no longer a buyer in the broad energy market; rather, it is a buyer in the narrower market for nuclear fuel. If, on the other hand, a utility builds a plant engineered to burn oil, gas, or coal, it may be able to convert to an alternative fuel at an affordable cost in response to changes in availability or price. Thus the fuel market may be defined broadly to include coal, oil, and gas.

Similarly, oil and natural gas are in the same market as a source of energy for heating, but not as a source of fuel for an automobile.

In antitrust cases, courts have adopted the same reasoning regarding the boundaries of a market. A famous case hinged, for example, on whether control of 75 percent of cellophane production constituted a monopoly of a narrowly defined cellophane market or whether cellophane should be considered part of a larger market for flexible packaging material. The court held that the relevant market was flexible wrapping material, and the du Pont company was acquitted.

In a suit against Alcoa, however, the court defined the market narrowly. It convicted Alcoa of monopolizing 90 percent of the production of new aluminum ingots, rejecting two other definitions of the market that would have shown Alcoa's market share to be either about 60 percent or 33 percent.

Geographic boundaries may also be taken into account by the courts. Firms that compete nationally may be found to be monopolies in certain regions or localities.

Although the principle of substitutability on which the definition of a market rests seems reasonably clear, the drawing of market boundaries proves very difficult in practice.

To operate efficiently, a market must provide information to buyers and sellers fully, accurately, and promptly. Traders need to know how much of a product is available, its quality, its price, and the amount of it that ultimate consumers are likely to want (hunches or forecasts of future trends also help). This information is essential in order for managers to make sensible decisions.

Thus, in major markets, newsletters, trade journals, and various long-range studies and projections supplement person-to-person conversations and meetings in disseminating information. But even so, much information is held closely within firms or among a few experts. In reality, information within the marketplace is typically partial, yet it may be good enough to let competition work.

If markets are to be efficient, they should also be open to new firms so that people with innovative ideas can get a chance to test them out and so that entrenched firms cannot coast along producing inferior products or restricting production to raise prices. The possibility that other firms may enter a market also influences the level of prices charged by established firms. Producers know that if prices are kept artificially high they run the risk of tempting new firms to enter and compete for a share of the business. On the other hand, when barriers to market entry are erected, say, by deliberately slow patent licensing, the power of established firms to intimidate potential competiton is enhanced and the effective and efficient operation of a market is inhibited because the market is insulated against competition.

The number and size of competing firms in a market is of special importance. Economists generally agree, on the basis of observation and study, that if a market consists of many buyers and sellers, no one of whom is powerful enough to affect the market price, and if they are all perfectly informed, and if entry into the market is not blocked by artificial means, then a market is probably performing its function efficiently—determining what should be produced, how much, and what prices should prevail. The functioning of some markets is also affected by environmental costs and other so-called externalities that affect outsiders who do not participate in the market transaction. In the economist's model of the perfectly competitive market, supply and demand—not the whim of a single producer or dominant buyer—guide production and prices.

In reality, markets differ from the competitive ideal in various degrees. In some energy markets, big suppliers can have an effect on price. In the world crude oil market, for example, when Libya came on the scene it produced enough to drive down the world oil price. Moreover, entry in many industrial markets is often difficult. It may cost too much. Essential technological knowledge may be protected by patents. Threats or agreements by firms already in the business may preclude a newcomer from developing customers. Access to an essential service (a pipeline, for example) may not be available. Furthermore, enough information may not be available to permit good decisions.

But if perfect markets do not exist, their attributes become useful in explaining the interplay of economic forces. As a pragmatic matter, economists often talk about "workable competition," or "competitive pressures"—terms signifying that there is enough rivalry and diffusion of power in a market to bring about results that are similar to those that would take place under perfectly competitive conditions. Deviations from the competitive ideal that appear contrary to the public interest may sometimes be checked through court action under the antitrust laws. If, however, a market cannot function effectively, for reasons discussed in the chapter on environment, or does not exist—as in the case of a natural monopoly—an alternative guide is government regulation, where price-setting is handled by a government body. The case for government regulation in a noncompetitive situation is not always wholly convincing. If a market is very small, if its transactions do not result in much redistribution of income, and the product or service is not essential, the benefits of regulation may be less than the costs. There is also a danger that regulation may help preserve the position of a monopoly that might otherwise collapse under changing conditions. But important areas remain where the alternative of regulation merits careful consideration.

THE WORLD OIL MARKET

The world market for energy is dominated by petroleum. Oil is the world's principal fuel; it is easily transported, and since production is concentrated in a few nations (many of them small and underdeveloped), most industrial nations are heavily dependent on supplies from abroad. Coal, on the other hand, is not only less heavily used, it is also more widely distributed throughout the world, and a higher proportion of countries can supply part or all of their needs. As for natural gas, trade is inhibited by the cost and difficulty of transporting it overseas.

During the 1970s the importance of oil in international trade tripled, and the structure of the world market was radically changed. By 1980, energy (primarily oil), accounted for 21 percent of the value of world trade, compared to 7 percent in 1970; oil alone made up 31 percent of all goods imported by the United States.

From Seven Sisters to One OPEC

Before 1973 the focus of the world oil trade was on the so-called seven sisters—Exxon, Mobil, Standard Oil of California (Chevron), Texaco, Royal Dutch Shell, British Petroleum, and Gulf. These giant multinational oil companies controlled almost all the important international reserves outside the United States and the Soviet Union. In most oil-rich countries, the government had long ago granted concessions to the companies, giving them exclusive rights to explore and produce oil within a defined area. The companies were, in effect, owners of the oil. They determined how much to produce, how much to pay to the foreign government and what price to charge for the final product. The major international oil companies effectively controlled the market in the non-Communist world.

The picture changed radically after 1973. Since then, governments of the oil-producing states have set production levels and prices. Their own national oil companies, along with companies owned by the governments of oil-importing nations (such as the British National Oil Company, Italy's ENI, and so on), have handled a large portion of international trade. Most of the oil in world trade has moved at prices based on, or linked to, those charged by the producers' international organization, OPEC.

Several factors contributed to this shift in the balance of power from the oil companies to the producing countries. The fundamental one was the growing world demand for oil. When there was excess capacity before 1973, the multinational companies were in a strong bargaining position. They could usually obtain supplies elsewhere if a particular government demanded an increase in royalties. With the decline in American production, however, and the increase in world demand, this was no longer true after 1970.

The power of the major companies was also eroded by the rising importance of the smaller "independent" companies, some of them owned by governments of industrialized importing countries. These companies, which were less likely to have several sources of supply, were more easily pressured by the producing countries than the majors, who had to worry that giving in to demands in one country would inevitably lead to calls for similar concessions by other governments. The state-owned companies were often even more willing to pay higher prices because the political cost of breaking established trade ties outweighed the cost of price increases.

A final factor was the rise of nationalism in the producing countries. More and more, governments were under pressure to reduce the power and influence of foreign companies, and they responded by nationalizing the oil companies in the early 1970s.

The 1973 Arab embargo served as the catalyst in changing the oil market. It irreversibly loosened the grip of the major companies. Led by Saudi Arabia, several producing countries issued orders for sharp cuts in production which the multinational companies obeyed. Subsequently, all OPEC members raised prices significantly. By the end of 1974 the price of crude oil had quadrupled over the period of a year, reaching nearly $12 per barrel.

The Oil Market in Recent Years

Since 1974, the producers have dominated the international market, and the influence of the largest multinational companies has been weakened. However, producing countries are not impervious to a decline in the world demand for oil, as the glut of 1981 and 1982 and the subsequent price cut has shown.

Of course, OPEC's price-setting ability is also limited by the different objectives of its diverse membership, and its price-setting meetings are sometimes acrimonious. Some members, such as Saudi Arabia, the United Arab Emirates, and Kuwait, have large oil reserves and small domestic populations.

They are comfortable with the prevailing level of revenues, and they can afford to take a moderate pricing stance. They tend to be more concerned about the long-term value of their oil. Others, such as Algeria, Nigeria, and Venezuela, have smaller reserves and large populations. They usually want to raise prices in order to meet pressing revenue needs and let others lower output as necessary to sustain the higher prices. Since their oil reserves may run out in twenty to twenty-five years, compared to Saudi Arabia's which are likely to last much longer, they do not worry excessively about the price of oil in the long term.

The market power of the biggest producer, Saudi Arabia, enforces a degree of stability on the OPEC members. The Saudis produced nearly half of OPEC's output in 1981 and, partly for that reason, their "Arabian light" crude oil is considered the OPEC reference point or "marker." That is, other crude oil is priced in relation to it. Crude oils from different wells are often of different qualities, with the lighter, low sulfur crudes of North Africa and the North Sea usually considered the best—that is, the easiest to refine into the most profitable mix of products. For this reason, producers of higher-quality oils charge a premium above the marker crude price. In practice, the higher quality of some oils makes them worth as much as 10 percent per barrel more than the Arabian light oil. Differences in location and transport costs also contribute to price differentials.

However, many OPEC countries, as a result of domestic political pressure, at times price oil significantly more than 10 percent per barrel above the Saudis. In November 1981, when OPEC official prices were allowed to range between $34 and $41 per barrel, prices ranged from $32 in Saudi Arabia to $40 in Algeria.

The increased influence of the government is evident in almost every producing country. Saudi Arabia, for example, controls vast reserves that were once the property of the Arabian American Oil Company (ARAMCO), a consortium comprised of Exxon, Mobil, Texaco, and Standard Oil of California. The Saudis tell the ARAMCO partners how much to produce and how much they will be able to buy for their own use. A government-owned company now sells a substantial portion of Saudi oil directly to other governments or companies. OPEC nations now own outright about 80 percent of their resources. Ownership of reserves by multinational companies outside the United States and Canada is primarily limited to areas such as the North Sea.

In addition, importing countries are also relying less on the multinationals. Feeling they may be able to make better deals by direct negotiation, governments of importing countries have also formed their own companies. These act largely as purchasing agents, although occasionally they participate in exploration and development of reserves, both at home and abroad.

Many Western European nations and Japan acquire well over one-quarter of their imports directly through government-owned companies or in deals in which the importing government is deeply involved. This percentage has increased rapidly in recent years, while the role of the multinational has declined.

Government-to-government deals introduced new political considerations into the oil market. Saudi Arabia has sold oil to Denmark on the condition that the Danes would not say anything the Saudis objected to about the political situation in the Middle East. Nigeria nationalized British Petroleum's operations because it objected to British policy in southern Africa. And industrialized countries have been asked to make substantial investments in producing countries as part of oil sales agreements.

The nature of international trade in oil has also changed in another way. The increasing number of players in the industry has led to an active open market for oil. Formerly, only small amounts of oil were traded openly because the major companies controlled the supply and refined and sold most of the oil they produced. Now more government-owned and smaller companies are involved. Many producers allocate their oil to centers of trading in Rotterdam, Singapore, and New York and sell to the highest bidder. A competitive market has arisen alongside the OPEC-dominated major oil market. Several companies and countries rely on this market for the bulk of their purchases.

Because of developments over the past decade, attempts to control or predict the behavior of the world oil market are exceedingly difficult. Following the cutoff of Iranian exports in 1979, the United States and other industrial countries made efforts to reduce the competition for oil on the "spot market"—a market where transactions not covered by long-term contracts are carried out—but could only watch in frustration as private companies bid up prices to unprecedented levels. Similarly, OPEC's attempts to unify prices have been frustrated by the ease with which member countries divert supplies to the spot market to get higher prices or offer relaxed credit terms to attract buyers when sales are slow.

Thus, though OPEC was the dominant influence in the world oil market for a decade, the effectiveness of its collective decisions was modified by defections of individual member countries, the rise of an international spot market, the persisting influence exerted by the multinationals, largely because of their expertise, and major shifts in world economic conditions, which sharply alter the demand for oil worldwide. In short, despite its success in driving up prices in the 1970s, OPEC was not all-powerful, and remained subject to inside and outside political and economic pressures.

These pressures affect not only OPEC's market power, but also its long-run viability. Historically, there have been attempts to form international cartels to control output and prices in many industries, such as chemicals and metals. The degree of success has varied. Held together by certain common concerns, countries participating in these agreements have also been pulled apart by differences in their own national interests. For a decade, OPEC nations maintained a strong cohesion—greater unity than many experts had pre-

From Wellhead to Market: Cost and Revenues of a Barrel of Imported Oil around 1980

A barrel of crude oil, which cost about 50 cents to produce in Saudi Arabia, yielded products worth nearly $50 at retail, of which 60 percent went to Saudi Arabia in rents and royalties, as shown below.

I. Costs

Cost of getting oil out of the ground in Saudi Arabia	$ .50
Rents and royalties to the Saudis	33.50
Cost of shipping to the United States	2.25
Cost to the U.S. refiner	36.25
Refinery costs	1.75
Estimated cost of refined products from one barrel of crude	38.00
Marketing costs, taxes, profit	11.19
Total	$49.19

II. Products from a barrel of oil

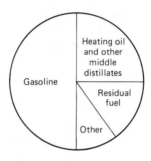

III. Revenues for refined products

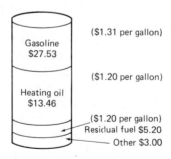

Source: These figures are approximations, since they vary with the quality of the crude oil and market conditions. Figures for production costs, shipping costs, refinery profits, and refinery output mix are estimated from government and industry data. Wholesale and retail prices are at 1981 levels and are from the U.S. Department of Energy's *1981 Annual Report to Congress* and the *Monthly Energy Review.* Revenues shown here are obtained by multiplying the barrel price of gasoline and other products by the percentage of each barrel of crude that is used for that refined product (for example, gasoline at $55.06 x 50 percent = $27.53).

dicted. How long they will continue to hold together, given the strains that have already been visible, is difficult to guess. The era of domination by multinational oil companies was replaced by the era of OPEC, and in the flow of history it would not be unusual for OPEC, too, to be succeeded by another marketing arrangement.

Meanwhile, OPEC continues to exist presumably because member countries are convinced that they can obtain a greater return on their production of petroleum with it than without it. No one knows precisely how much higher prices are, and how much lower production is, because of OPEC. No one knows what the level of output and prices would be today if oil could be produced and marketed under the conditions of a truly competitive market (a type of world market, incidentally, that has never existed for oil). But no one doubts that in the 1970s OPEC succeeded in dominating the world oil market, sharply increasing the revenues of member countries.

The structure and organization of the world oil market contrasts significantly with that of the domestic U.S. market. At least since 1911, when the Standard Oil trust was broken up, American economic policy has reflected a belief that, to the extent possible, the marketplace should be governed by competition among independent companies rather than by the dictates of a monopolist or by a group of firms collaborating to further their special interests. Government intervention in energy markets, as discussed below, has at times been inconsistent with this view. Still, a competitive market is the type of market preferred in the United States because of the belief that it can lead to greater efficiency. That this goal has been only partially reached has not greatly weakened the attractiveness of the ideal.

COMPETITION IN THE U.S. PETROLEUM MARKET

How can we determine whether markets are competitive—whether output and prices are being guided largely by impersonal forces of supply and demand rather than by a dominant firm or group of firms? Extreme cases present no problem. They are readily identifiable. OPEC and a local utility make no claim to being competitive institutions. The more typical situations, however, are ambiguous. Elements of monopoly and competition exist side by side.

Unfortunately, there is no simple litmus test that will signal either "competition" or "monopoly." But some indicators have been devised that can assist us in drawing inferences about the state of competition in a market. They are related to three aspects of the market: the structure of the industry, the conduct of the firms that comprise it, and the performance of the industry compared with that of other industries. Each of these measures tells something about the nature of a market, and collectively they can tell a good deal. They are described below as they apply to the U.S. oil market.

Market Structure, Conduct, and Performance

Structure.   The structure of an industry refers principally to the number of firms in it, their relative size, and the obstacles to entry by new firms. The market for crude petroleum can be used for illustrative purposes. Casual acquaintance with the industry suggests that large firms play a big role, but one needs to probe more deeply into the matter. One step toward greater precision is the analysis of "concentration ratios," which measure the percentage of output of an industry that is dominated by big firms—usually the four largest firms, or the eight largest, or twenty largest. As a rule of thumb, for example, if the four largest firms account for 70 percent of sales in an industry, there is reason to be worried about the possibility of collusion—that firms can get together to limit output, set prices, and so on, in a way that will be most advantageous to themselves. Concentration ratios thus are useful as a screening device for identifying potential problem areas. Other types of analyses might indicate, however, that competition is vigorous in an industry dominated by two firms and weak in an industry consisting of dozens of firms about the same size.

Table 5-1 shows concentration ratios for the U.S. crude oil industry in comparison with a number of other industries that are measured by the U.S. Federal Trade Commission. It appears from the data—broadly representative of the range of FTC figures—that production is less highly concentrated in the oil industry than in many other industries, and, on the basis of this measure, there is no reason to believe that the U.S. crude oil market is relatively less competitive than other American markets. It should be pointed out, however, that these concentration ratios are based on the assumption that the U.S. market is the relevant geographic boundary for crude oil. But as discussed above, crude oil is traded internationally and the concentration ratios would have to be adjusted accordingly in order to measure more accurately the actual degree of market concentration. Nonetheless, even this preliminary look at concentration ratios illustrates the difference between the absolute and relative size of firms. Measured by their assets or profits, the major oil firms are indeed giants, yet their share of the market is no greater than that of far smaller firms competing in smaller industries.

One of the limitations of the concept of concentration ratios is that it does not take into account the extent to which firms in the same industry differ from each other, especially with regard to their operations in the several submarkets that typically comprise contemporary industries. Concentration ratios in oil refining, for example, do not take into account differences in the refiners' market positions. Some refineries are independent firms that neither own sources of supply nor distribution outlets; they are solely refiners. Other refineries are operated by firms that also own oil reserves and that can supply fully the needs of their refineries. Still others may own refineries and distribution outlets. To focus solely on concentration in the refining market, for exam-

TABLE 5-1.   Concentration Ratios in Selected U.S. Industries, 1954 and 1977

Industry	Percentage of shipments by four largest firms		Percentage of shipments by eight largest firms	
	1954	*1977*	*1954*	*1977*
Blast furnaces and steel mills	55	45	71	65
Cereal breakfast foods	88	89	95	98
Chewing gum	86	93	94	99
Coal production[b]	17.8	21.1	25.5	29.5
Crude oil production[b]	18.1	25.3	30.3	40.8
Electronic computing equipment	66[c]	44	83	55
Greeting card publishing	42	77	57	85
Milk	22	18	28	28
Motor vehicles and car bodies	92[c]	93	98.6	99
Natural gas production[b]	21.7	19.2	33.1	30.6
Petroleum refining[a]	34.7	29.0	57.9	49.0

Note: The percentages are calculated on the total dollar value of an industry's shipments; n. a. indicates data are not available. Energy industries appear in *italics*.

Source: Data are taken from *Concentration Ratios in Manufacturing,* U.S. Bureau of the Census, 1977 Census of Manufactures, May 1981; and *Market Shares and Individual Company Data for U.S. Energy Markets: 1950–1980,* Amercian Petroleum Institute, October 1981.

[a] Figures are for 1950 and 1980.

[b] Figures are for 1955 and 1980.

[c] Figure is for 1967.

ple, as a measure of the degree of competition in an industry actually comprised of several vertically related markets, is to overlook the market advantages conferred on firms that own their own supplies of petroleum and control their own distribution organizations. Therefore, one certainly is warranted in asking the question whether vertically integrated firms—those that compete in extraction, transport, refining, and distribution of petroleum products (that is, firms that confront each other in several markets)—behave differently from firms that operate in a single market—refining, for example.

One recent study of multimarket industrial structure, which focused on the petroleum industry as a case study, suggests that vertically integrated firms may be less competitive than nonintegrated ones. Integrated firms, according to this analysis, tend to make larger profits that are attributable to their market power, that is, the ability to charge prices in excess of cost. Firms facing each other in a series of vertically "linked" markets may hesitate to pursue aggressively a competitive advantage in one area lest a competing firm push its advantage vigorously in another area. This finding suggests the need to supplement conventional measures of market competition, such as concentration ratios, with other measures that account for the structural heterogeneity of firms within an industry.

Conduct. Aside from structural conditions, the abuse of market power within an industry may be reflected in the conduct of aggressive firms. Do they try to drive competitors out of the market through unfair practices? Do they act illegally to prevent newcomers from entering the market? Do they make agreements among each other—or tacitly follow an industry leader—to control output and prices?

There is little evidence to indicate that there is unchecked, highly anticompetitive behavior within the oil industry today—certainly nothing comparable to predatory behavior of the old Standard Oil Company, which was dissolved by the famous Supreme Court decree of 1911, or to the kind of collusion that led to the conviction of the electrical equipment companies for price fixing in the 1960s.

One aspect of business conduct, however, repeatedly arouses concern—the takeover of smaller firms by giants in the industry, or the combination of two large firms to increase their market power. In part, the concern over mergers merely reflects a historical public mistrust of bigness—in government and in business. More important, it reflects the fear that certain mergers may weaken competition and lead to higher prices.

These concerns are embodied in the antitrust legislation dating back nearly a century that has sought to preserve competition in the U.S. economy. The Sherman Antitrust Act of 1890 forbade agreements among companies in restraint of trade and efforts to monopolize the market. The Federal Trade Commission Act and the Clayton Act of 1914 and the Antimerger law of 1950 modified and carried the policy forward, though the vigor of enforcement by the Justice Department and the FTC has varied over the years. Moreover, a number of significant court decisions over the years defined and reinterpreted the vague language of the antitrust statutes. The Supreme Court has held, for example, that bigness in itself is not a violation of the law. Mergers to achieve vertical integration have usually not been opposed by antitrust authorities; they may attract governmental scrutiny to make sure that they are not likely to lead to a substantial lessening of competition, but, in general, such mergers have not been found to be illegal. On the other hand, horizontal mergers—combinations of firms that compete in the same market—are likely to be challenged by the government since they reduce the number of competitors.

In the petroleum industry, much of the vertical integration has been achieved by mergers. To assure themselves of a supply of crude oil, refiners buy producing companies; and refiners also buy marketing companies to control the distribution of their output. Even large vertically integrated companies may merge in order to obtain a better balance of resources at each stage of production. Although, in principle, vertical integration is accepted, questions may be raised if a firm's behavior in seeking to extend vertical integration appears to reflect an abuse of economic power and an intention to weaken competition.

Other types of mergers have also taken place. Acquisitions of coal companies by oil firms has also been widely permitted even though in some product markets the two fuels are competitive.

Some mergers have both vertical and horizontal aspects. For example, Mobil, the second largest oil company, encountered a legal snag in late 1981 when it sought to buy Marathon Oil, the seventeenth largest. Mobil, which produces only about a third of the crude oil that it needs for its refineries, said the merger was intended to strengthen its holdings of oil and gas reserves. The court held, however, that the effect of the merger would seriously weaken competition in the Midwest area where Marathon was a major distributor of oil products. Mobil offered to sell off Marathon's Midwest oil-distribution system, but the court blocked the acquisition.

Over the years, the antitrust laws have been the subject of much debate. The necessity for them and their effectiveness has been questioned. The issues are complex and controversial, but there seems no doubt that the laws will stay on the books and that firms may be discouraged from merging and engaging in anticompetitive actions—or prevented from doing so—in the name of protecting competition.

Firms are allowed some latitude in other activities that may affect competition. Even within a competitive market, certain collaborative activities are permissible because they ostensibly are separable from the production and sales activities in which competitors engage. Competing firms are permitted, for example, to form "joint ventures"—to work together in a new activity—on grounds that some ventures are too risky or too costly for a single firm to undertake on its own. They are widespread especially in two fields: pipeline operations and joint bids to lease tracts for oil exploration.

In the pipeline segment of the industry, the reasons are straightforward and unexceptional. Joint ventures are dictated by economies of scale. A single large pipeline is much more economical than several small competing ones, and government regulations are designed to assure all firms equal access to the pipelines at fair prices.

In leasing, the situation is less clear cut. Given the large risks associated with oil exploration, firms can make a good case for spreading their investment among several leases rather than putting all of their money into a single one, or a few. Moreover, the cost of leasing and exploration has increased substantially as the search for new oil fields has led to drilling at greater depths and to a higher proportion of drilling offshore. Thus, if they were unable to collaborate in leasing, many medium and small firms might not be able to participate actively in the search for new oil and gas reserves.

On the other hand, joint bidding by firms may reduce the government's return on leases by reducing the number of bidders and perhaps the size of bids on federal tracts. Moreover, the need for joint bidding is questionable in some instances. The biggest firms are quite capable financially of bidding for leases without the resources of a partner. For example, Exxon, the biggest firm, independently owned four out of five of its leases in the mid-1970s, and Standard Oil of California, the fifth largest firm, independently owned four out of five of its leases. (Mobil, then the fourth largest, independently owned slightly over 10 percent of its leases.)

Some of the concern about joint bidding was eased in the 1970s when the U.S. Department of the Interior banned joint bidding by a handful of the largest companies (those with worldwide production in excess of 1.6 million barrels a day).

Performance.   Two yardsticks are commonly used for measuring performance—(1) rates of profit, and (2) expenditures for research and development. If profit rates are excessive, one may want to look more closely to determine whether they are attributable to market power rather than to expanding markets, rapid innovation, and efficient management. On the other hand, market power may not necessarily result in an excessively high rate of profit, and so the profit measure is not always a reliable indicator. A major goal of the monopolist, as has often been cited, may simply be to enjoy the quiet life. Lack of competition may enable a poorly managed firm or group of firms to survive with average rates of profit despite a substandard product. A low rate of earnings may suggest poor management in a monopolistic market rather than vigorous rivalry in a competitive one.

Earnings data for the petroleum industry show that the performance of the giant firms is above the median for all firms but falls in the range of successful firms in U.S. industry in general. Profit rates of the internationals, measured as a percentage of capital investment, were generally below those of such firms as CBS, General Motors, and IBM, and about the same as for McDonalds, and way above that for U.S. Steel, as shown in Table 5-2. Until the embargo of 1973, the median profit rate of the energy industry was the same as the return on total capital for all U.S. industries. For the 1971-75 and 1976-81 periods, however, profits of the international oil companies were substantially higher. Clearly, the 1970s was a good decade for them, but profits dropped sharply in 1982, and prospects for the decade were not as bright as the performance of the 1970s.

As for technological performance, judgments are difficult to make. Expenditures for R&D by large petroleum firms seem to be in line with those by large firms in other industries, but there is no evidence to indicate that large firms spend proportionately more or achieve greater inventive success than small ones in the petroleum industry, once a certain minimum size required for R&D efficiency has been met. Of course, R&D expenditures do not in themselves show what kinds of research are being supported, how successful the research is, and how important are the innovations that result. Some studies of this issue suggest that large oil firms have tended to concentrate on

TABLE 5-2. Average Annual Profit Rates of Firms in the Oil Industry and Other Industries for Selected Periods (percentage return on total capital)

Firm [a]	1966–70	1971–75	1976–80
CBS	14.9	15.5	19.6
Coca-Cola	26.2	23.2	22.3
Conoco [a]	9.0	10.2	14.4
Exxon	11.5	14.1	13.8
General Motors	17.4	14.8	15.1
Gulf Oil	11.1	9.8	11.4
IBM	18.2	18.2	21.5
McDonalds	n.a.	15.1	13.6
Mobil	9.5	11.5	13.2
Occidental Petroleum	11.9	6.8	14.0
Phillips	7.3	9.7	16.5
Standard Oil California	10.5	11.9	14.4
Texaco	13.2	11.6	10.4
United States Steel	5.1	7.5	1.3
Wang Laboratories	n.a.	n.a.	17.7
Energy industry median	9.5	11.2	13.3
International oils	n.a.	10.9	13.9
Other oil and gas	n.a.	11.0	13.6
Coal	n.a.	15.6	8.5
All industries	9.5	8.6	11.1

Note: n.a. indicates data are not available. Oil companies appear in italics.

Source: "28th Annual Report on American Industry," Forbes, January 1, 1976; "23rd Report on American Industry," Forbes, January 1, 1971; and "33rd Report on American Industry," Forbes, January 5, 1981.

[a] Taken over by Dupont, 1981.

relatively safe R&D that is likely to pay off within five years, but that the top four may have put relatively greater emphasis on basic research. Historically, independent inventors have been responsible for major breakthroughs in refining. However, these studies do not include recent data and the conclusions are not clear cut.

Summing Up

Our examination of structure, conduct, and performance uncovers no serious causes for concern, but neither does it prove that the petroleum industry in the United States is without competitive problems. What it does suggest is that the measures that economists have so far developed to assess competitiveness are only rough indicators of a very complex problem. Thus it is very difficult to derive conclusive answers about many of the questions that concern people—for example: is the world oil market governed by supply and demand or by OPEC? We saw that when world demand for oil was high relative to the supply, OPEC was able to raise prices dramatically. But when demand fell without an accompanying drop in supply, OPEC was unable to hold the line on prices. One might reasonably infer that in the world market, supply and demand set boundaries within which OPEC can maneuver. But should such a market be described as essentially competitive or monopolistic? Perhaps neither label should be considered appropriate.

In the United States market, too, we have seen that the relative strength of market forces and individual giant firms is not easily determined. Moreover, conventional economic analysis does not even attempt to deal with another dimension of economic power, its political implications. Discussions of public policy deal not only with beliefs about whether big firms can determine how much shall be produced and what price shall be charged, but also with views about the influence of big firms on legislative and administrative decisions and on public attitudes through advertising and public relations activities. Within the framework of this book we can do no more than call attention to this question and recognize the limitations it may impose on inferences drawn from our analysis.

Regulation of Oil and Gas

For a half-century competition in the oil and gas industry has been modified by government policy or replaced by regulation. Thus the heavy intervention of the government after the oil embargo of 1973–74 can hardly be viewed as a sharp departure from the past.

Two examples serve as strong reminders of this experience. The U.S. government has displayed a visible hand since the 1930s. First, during the early 1930s when the demand for oil was weak, the oil- and gas-producing states, led by Texas and with the blessing of the federal government, intervened under the guise of "conservation" to control oil production. In effect, a system of state-controlled cartels was established that kept prices from falling by adjusting the output of oil in line with expected changes in the demand for it. In retrospect, it can be said that the objectives of the states look somewhat similar to those adopted later by OPEC. Second, in 1959 the United States imposed a system of import quotas because the world price of oil had fallen below the level of U.S. prices. The quota made it possible to sustain the price of domestic oil at $1.00 to $1.50 a barrel above that of imported oil.

Both of these steps—one by states and one by the federal government—served to protect domestic producers from price competition. At the same time they imposed a higher price on business and household consumers of petroleum products and encouraged the depletion of U.S. reserves.

The regulations imposed after the OPEC oil embargo, however, sought to benefit consumers. After crude oil prices quadrupled in the 1973–74 period, the government acted to protect consumers (and some sectors of the oil industry, particularly independent refiners) from bearing the full brunt of the high prices imposed by OPEC. Special regulations were established to help users in the northeast states, who were especially dependent on oil and vulnerable to price increases. In addition, a "two-tier" system of price controls was imposed. One governed the production of "old oil" from existing wells, which was kept at a lower price to keep windfall profits in check. "New oil" from newly drilled wells (or additional oil produced from old wells operating in excess of previous rates of production) was priced higher. An elaborate entitlements program was set up to equalize crude oil costs between refineries that were primarily dependent on expensive foreign oil and refineries that depended mostly on cheaper, price-regulated domestic oil. Over time, provisions of these regulations were changed many times by congressional, administrative, and executive actions to deal with criticisms of their effects on inflation, on imports, and on the incentives of oil producers.

The move toward deregulation of oil and gas in the late 1970s reflected growing concern about the consequences of government regulations. A study financed by the Ford Foundation, for example, put the yearly cost of government energy regulations at $3 billion. The precise dollar amount of the estimate is less important than the analysis showing how these costs arose from the following kinds of undesirable effects.

First, by holding down the price of oil, government policy served to encourage greater oil consumption, and as a result U.S. imports continued to be higher than they would have been otherwise. Second, the lid on prices reduced the incentive for oil firms to search for and produce more oil. In particular, oil that was costlier to produce because wells were deeper or less productive, was kept off the market. The estimated cost of losses in efficiency from consuming more and producing less was placed at $1 to $2 billion in 1978. The government's costs of administering the regulations and the costs imposed on firms that had to comply with regulations were put at $750 million. There were also nonmonetary costs, such as the inconvenience of waiting in line for gasoline, and the impairment of the economy's capacity to adjust to world energy prices. Moreover, uncertainty about future changes in government policy disrupted long-standing ways of doing business.

In short, the imposition of controls in the 1970s changed the rules of the game, protecting consumers and some refiners while limiting the profits of big oil companies. Many observers had argued that the nation would have been better off if pricing and equity issues had been addressed separately—if prices had been allowed to rise and then steps were taken to compensate for their impact. Taxes might be imposed on domestic oil producers, who realized a windfall gain on their assets merely because of the foreign price hikes, for example, and financial aid provided for those who were hardest hit by the price increases. There is little evidence to suggest that the regulations were essential or, on balance, helpful, though they did help to avert a major confrontation over remedial measures to equalize burdens on consumers and oil producers.

Natural gas, often jointly produced with crude oil, also has had a long regulatory history. In 1938 passage of the Natural Gas Act gave the federal government the power to regulate the price that could be charged for pipeline *transmission* of gas in interstate commerce. In 1954 a Supreme Court decision in a dispute between the state of Wisconsin and the Phillips Petroleum Company made the *production* of gas destined for interstate sale also subject to price regulation.

The case that has been made for regulating the pipeline transmission of natural gas to utilities and local distribution to users is the familiar one made for a natural monopoly—that is, an industry in which costs continue to decline as the size of an operation increases, with the result that costs will be higher if the service is performed by more than one company. However, critics of regulation point out that this argument does not apply to the production of gas and, moreover, production is no more concentrated among a few firms in the gas industry than in most industries (see Table 5-1).

After the early 1970s, government regulation kept the price increases of natural gas sold in interstate commerce below the price increases for intrastate gas—gas produced and used within the same state. By 1975, as a result of a strong demand for natural gas and occasional shortages, unregulated prices were about 50 percent higher at the wellhead than were regulated prices. Producers, as a result, channeled most of the additional gas they produced into the intrastate market. Consumers of cheap, regulated gas had little incentive to conserve; and they doubtless consumed more gas than they would have if prices had reflected actual market conditions.

The effect of this regulatory policy was to transfer income from producers to those consumers who were fortunate enough to obtain the gas they needed. Moreover, since there was not enough gas to go around at the regulated price, other prospective buyers were shut out of the market. Thus some consumers had to rely on higher-cost oil while others benefited from cheap natural gas.

The case for decontrolling natural gas gained adherents through the 1970s, and, in 1978, the Natural Gas Policy Act was passed to raise gas prices gradually toward the market level. The goal was complete deregulation of new gas in the 1985-87 period, and it was to be achieved gradually in order to cushion the shock of higher prices. Old gas would remain regulated, however, with the result that the average price of gas would be lower than a free market price. At the same time, intrastate gas was placed under regulation as part of the effort to eliminate the incentive to channel gas into the intrastate market where higher prices had prevailed. Prices rose more rapidly than expected, however, and by early 1983 the real cost of gas to residential consumers was 50 percent higher than it had been in 1978. The burden fell disproportionately on the poor, especially in cold regions. Emergency relief was provided in some areas, and policies to ease the transition came under active discussion.

The long history of government efforts at regulation raises the question whether the recent swing toward deregulation is a temporary phenomenon or an enduring change. Will producers and consumers be willing to abide by the results of competitive markets, or will either group, or both groups, again turn to government for support if unfavorable events should occur?

THE COAL INDUSTRY

The structure of the coal industry was transformed during the 1960s and 1970s. Coal production, which was long dominated by companies engaged solely in the coal business—the so-called independents—has increasingly come under the control of firms from other industries. In 1960 twenty of the top twenty-five coal-producing companies were independents. By 1975, there were only four independents in the top twenty-five. The other major coal producers were subsidiaries of larger firms. Nine were owned by other energy companies (including Exxon, Gulf, Occidental, Sohio, and Ashland); six were the so-called captives of the steel industry or electrical utilities, and six were controlled by conglomerates with an interest in developing synthetic fuels. The leading producer, which had long been the Peabody Coal Company, had become the huge Peabody Holding Company, a giant diversified firm.

A significant proportion of coal production—an average of about 20 percent over the years before World War II—had long been owned by the steel, utilities, and railroad companies. These so-called captive mines represented a kind of vertical integration by the consuming industries, which wanted to be assured of a stable source of supply to meet their needs. However, the use of coal by railroads declined sharply after World War II as trains began switching to more efficient diesel engines; it virtually ended by the mid-1960s, though railroads continue to be major owners of coal reserves. Residential and commercial use also dropped sharply during the period. But these losses in rail and other uses were more than offset by a doubling of consumption by the electrical utilities industry. Utilities now depend on coal for about half of their energy needs, and as their dependence on coal increased, they bought more coal companies and became the dominant owners of captive mines.

The purchase of coal companies by oil and gas firms has been partly a hedge against eventual depletion of oil and gas supplies and partly an effort to get a foothold in the potential synthetic fuels industry. The entry of conglomerates, however, is a new trend. The conglomerates that moved in—largely firms at the forefront of technological development—were seeking to assure themselves of a source of supply if their research and development of synfuels paid off.

Although the toll among independents during the 1960s and 1970s was heavy, numerous small independent operators remained in business. Historically, the independents had moved in and out of the market as demand for coal increased or declined. They survived because coal deposits and pro-

How Energy Is Transported*

Moving energy supplies from where they are to where they are needed requires a variety of transport modes—barge, truck, railroad—to fit the characteristics of the fuel—liquid, gas or solid—and geographic factors. Transport is also a major element of cost—accounting for about a third of the cost of natural gas, for example.

Natural gas

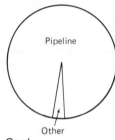

Natural gas presents the simplest case. Virtually all of it moves by pipeline. No alternative mode can match its advantages. Because of the efficiency of large-scale operations (pipe diameters have been increased to 48 inches), pipelines are generally monopolies regulated by the government. Most of the gas flows out of Texas, Louisiana, and contiguous states to the rest of the country, except for the Pacific Coast, which is supplied by California.

Coal

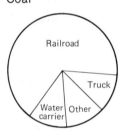

Coal is largely a regional fuel, even though minable deposits exist in thirty states. Four of the five largest producing states are among the five largest consuming states—Ohio, Pennsylvania, West Virginia, and Illinois—and they are all in the Great Lakes–Appalachian region. Most of the coal is moved by rail.

Crude oil

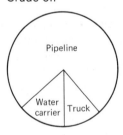

About 75 percent of the crude oil produced in the United States moves from wellhead to refinery by pipeline; the rest is divided between water carriers and trucks.

Refined products

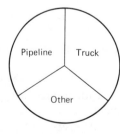

Gasoline and other refined products move from the refinery to consumers primarily by truck and pipeline. Nearly half of the refining takes place in the Gulf region near the producing areas; a fourth is handled in the Great Lakes and mid-Atlantic region, and most of the remainder in California. (Because of the relatively small volume, the location of sites involved in the nuclear fuel cycle, and safety aspects, almost all of the nuclear traffic is handled by truck.)

*This material is taken from the U.S. Senate Committee on Energy and Natural Resources and the Committee on Commerce, Science and Transportation, *National Energy Transportation*, vol. 1, GPO Publication No. 95–15 (Washington, D.C., GPO, May 1977).

duction are scattered over half the states in the United States, and there were opportunities to serve local markets. The growth of mechanization, as well as the merger trend, reduced their number to about 6,000 in the 1970s, but still within the historical range of 5,000 to 9,000.

The shuffling in ownership and control has not led, however, to an alarming rise in concentration statistics. The top four coal producers controlled about 21 percent of the market in 1977, up from about 18 percent in 1954, but concentration remained lower than in oil production (see Table 5-1). The top twenty-five coal companies controlled about half the production. Neither of these measures suggests the presence of unusual market power.

The market for coal is also affected by the fact that coal is less homogeneous than oil. While variations in the characteristics of crude lead to a range in price per barrel, the variations in the quality of coal—especially in sulfur content, which is important for environmental reasons, and in the amount of heat that will be obtained per ton—lead to a wider spread in prices per ton. Coal sales may also be affected by changes in oil prices. Geographic factors, the location of mines, and the location of buyers affect transport costs.

By and large, those engaged in transporting coal to markets are independent of the producers. Two-thirds of the coal moves by rail and the rest by barge and truck. Pipelines, which are important in oil and gas and controlled by producers, play an insignificant role in coal transport. There has been continuing interest in slurry pipelines—which would carry crushed coal combined with water—especially from western coal areas, but local opposition based on scarcity of water and the withholding of rights-of-way by railroads has kept them from being built. The railroads have thus been able to raise a barrier to entry and thereby prevent the development of competition in coal transport.

The coal market has also been altered by the decline in the relative importance of underground mining. Even in some of the eastern and midwestern states, more coal is now produced by strip mining than by underground mining.

What do these changes in the coal industry portend? For the foreseeable future, the utilities will continue to be the biggest coal consumers, followed by the domestic steel industry and the export market. By 2000 and beyond, however, if R&D in synthetic fuels leads to a synfuels industry (see Chapter 4) and prices for oil continue to rise, prospects for a broader coal market look promising. The acquisition of coal production by high-technology firms can be viewed as one indication that synthetic fuels may eventually become very important, though enthusiasm rises and falls with changes in oil prices. The acquisition of independent coal producers by large and economically powerful firms in the oil industry and in coal-consuming industries, may act to stimulate research and development. Not only may more money be spent on research, but the emphasis will shift away from a narrow concern for improving coal extraction and transportation to the development of new uses. As for the structure of the industry, there would seem to be sufficient diversity in both production and consumption to generate effective competitive pressures. As yet, there are no sizable artificial barriers to entry into the coal market. Moreover, for the near future at least, other fuels remain competitive with coal in many uses.

ELECTRIC UTILITIES

Electric power, which accounts for more than one-fifth of the energy used by households and commercial firms in the United States, is seldom available in a competitive market. It is supplied by local monopolies that are regulated by government agencies or publicly owned. Three out of four of these electric utilities are owned by private investors, and the rest by municipalities and other governmental units, or by rural cooperatives.

The private power firms are generally integrated. They generate electricity, transmit it, and distribute it to industry, commerce, and consumers. Some control their source of coal supply.

The rural cooperatives, however, typically perform only one of the three functions; they are primarily distributors. First organized in the mid-1930s with support of the federal government, they have brought electricity to sparsely populated areas that initially did not offer attractive investment opportunities to private firms. Today they provide power to 25 million consumers, mostly in communities with populations under 10,000. They obtain their power in roughly equal proportions from private companies, the federal government, and from those cooperatives that generate as well as distribute power.

The federal government, on the other hand, is primarily a generator and wholesaler of electricity at such facilities as those run by the Tennessee Valley Authority and the Bonneville Power Administration; fewer than 1 percent of users get their power directly from the federal government. Initially the federal facilities were established to capitalize on hydropower possibilities on the Tennessee and Columbia rivers and to establish a yardstick for the pricing of electric power. In recent years they have been augmented by coal-burning and nuclear plants. Almost all of federal electricity is sold at wholesale to other utilities and to federal installations rather than to consumers.

Municipal power plants—which provide electricity to a few large cities such as Los Angeles and Seattle but primarily to small ones such as Severance, Kansas, and Albion, Idaho—generate about half of the electricity that they distribute, and buy the rest.

This mixture of investor ownership and public ownership (local and federal) is unusual in the United States. No division of ownership on a comparable scale exists in U.S. industry though consumer cooperatives are well established, especially in agriculture, and municipal ownership of water, sewage, and transportation facilities is common.

The industry can be described by function as well as by ownership. Thus an estimated 53 percent of the costs to the final user occur in the electricity-generating sector, 12 percent in transmission, and 35 percent in distribution.

For the nation as a whole, coal is the principal fuel used in the generation of electricity. Along the Pacific Coast, however, hydropower has predominated, and in Texas, Louisiana, and contiguous states, gas has been the principal fuel. The Mountain States use about equal amounts of hydropower, coal, and gas. Oil is an important source in the eastern states, primarily in New England and Florida.

These differences reflect to a large extent geographic advantages. Hydropower is completely dependent on proximity to powerful rivers. The use of natural gas has been encouraged in the Southwest where, as a result of federal regulatory policy, there was an incentive for producers to sell gas in the state where it was produced rather than outside the state. Elsewhere, relative prices, including transportation costs, have generally been decisive. Some areas enjoy flexibility. About one-third of the nation's power plants can be adapted to burn either coal and oil, or coal and gas depending on the availability of supply and the prices of the alternative fuels.

Pricing Practices

The principal regulatory problem for state utility commissions is the setting of prices. Price setting deals not only with the level of prices but with the structure as well—that is, with the treatment of different classes of buyers, commercial and residential consumers. For investor-owned companies the price issue is inextricably tied to the rate of return—How much should investors earn? In some municipalities, price-setting and other managerial decisions are made by locally elected commissions, but the bulk of utility regulation is conducted at the state level, not only with regard to prices, but also for compliance with environmental requirements and other regulations.

The objectives of the state utilities commissions are to make sure that there is an uninterrupted supply of electricity at appropriate rates. For privately owned firms, this means that the rates are generally set to yield a sufficient return to attract enough funds from investors to maintain and improve power generating and distribution facilities. Commissioners are also subject to political pressures from consumers and companies, and rate-setting is not simply a technical exercise.

Until the 1960s, regulatory agencies pursued a promotional price policy that was intended to increase the consumption of electricity. Increases in consumption led to greater utilization of existing facilities and the construction of larger plants that were more efficient and capable of turning out more electricity per dollar of investment. The success of this policy helped encourage the extension of electricity to all parts of the United States, even for the smallest rural communities. As consumption increased, rates fell.

Rationing

During World War II, gasoline was rationed and its price was controlled as part of the nationwide effort to meet military needs with a minimal disruption of the civilian economy. During the oil supply shocks of 1973–74 and 1979 standby rationing plans were considered though they were not imposed; nevertheless, government played a big role in energy distribution through allocations of petroleum supplies and price regulations. In general, when there is a threat of a sudden and substantial gap between the supply of an essential product and the demand for it at the prevailing price, there is a great pressure for government action of some sort to ease the shock and equalize hardships. The question is, What is most appropriate and effective?

Two variations of the approaches used in the past have been widely discussed in recent years.

Under the free-market approach, the government would permit prices to rise so that the market could continue to play its role in allocating supplies and facilitating transactions; meanwhile, a windfall profits tax would be imposed to prevent the energy industry from benefiting unduly from the emergency; tax revenues could be distributed to the public or used to finance other public programs.

The second approach would be to impose rationing. One much-discussed variety, aimed at making the system more flexible and fair, would establish a market for coupons so that they could be bought and sold openly.

The value of marketable coupons would be determined by the market. It would be equal to the difference between the controlled price of gasoline—say $1.50 a gallon—and the price that would prevail if there were no price controls—perhaps $3. Motorists who chose to use their full ration would be able to get their quota of gasoline at no additional out of pocket cost per gallon. Those who chose to drive less could sell their coupons for cash. Those who need to drive more could get additional gas by buying coupons. The real price of gasoline under this system would be the pump price plus the value of the coupon, the same price that would prevail in a free market.

No system of rationing, allocation, or free market interplay, however, can eliminate the fact that a sharp curtailment of supply imposes a cost on society. Public policy can seek only to minimize the cost by seeing that essential needs are met (military, food supplies, medical care, and others) that businesses and individuals are treated equitably, and that the normal operation of the economy is disrupted as little as possible.

One of the instruments devised to encourage greater electricity consumption was called the "declining block" rate, which provided that as users increased their consumption by successively higher amounts, they paid less per kilowatt for what they consumed.

In recent years several factors have led to a questioning of this approach to pricing. The most important one is that utilities no longer can reduce their generating costs by building bigger plants. They appear to have achieved the feasible "economies of scale." Some analysts believe that plants are most efficient at a size of 400 or 500 megawatts (MW). (A 500-MW plant would serve a community of about 500,000 people.) During the 1970s, rapidly rising prices and soaring interest rates greatly increased the cost of construction and threatened profitable operations generally. Policies to protect the environment have imposed additional costs. As a result, companies no longer emphasize incentives to use more electricity. The theme of public policy has been to use electricity more efficiently and thereby reduce the consumption of fuels, especially fossil fuels, which in turn will reduce pollution. (To encourage conservation, the federal government has offered tax breaks to homeowners for investments in insulation, and some states have required utilities to lend money at low rates to homeowners for conservation investments.) The rate of increase in the construction of new plants has declined in some areas, and in others the need for additional plants has been eliminated. The shift from promoting the consumption of electricity to conserving it will be hindered, however, so long as the old approach to rate-setting prevails. A different approach is needed that takes into account the higher costs of using additional electricity. Successive megawatts of electricity cost more when they can be provided only beyond their optimum efficiency by drawing on less-efficient standby plants or by constructing more costly new plants.

Pricing Alternatives

One of the alternative pricing procedures that has attracted attention since the mid-1970s—when test cases were won by state utility commissions in California, Michigan, and Wisconsin—is so-called time-of-day (or peak load) pricing. This approach recognizes that (1) the need for plant capacity is determined by the maximum amount of electricity that must be generated at any one time, and (2) the peak load is determined by the season and time of day (in addition to general economic conditions). The demand for power in most areas is greatest during winter's coldest times and summer's hottest days. Demand is also higher during the daytime hours when factories are working full blast, businesses are open, and consumers are busiest. If these variations can be evened out, so that the peak load is reduced substantially, the use of less efficient facilities can be reduced and investment in new plant and equipment postponed.

Peak load pricing of electricity—like low night and weekend rates for long-distance telephone service—is an attempt to achieve that result. It offers lower rates for electricity consumption in quiet seasons of the year and in the evening and early morning hours of the day. The lower rates are intended to be enough of a bargain to get customers to alter their patterns of use. For the electric companies, the loss of revenue from rate reduction is intended to be more than offset by the higher rates charged at peak time and by savings in operating costs and postponement of capital investments. It is still too early to tell whether this approach will become widely accepted.

Electric utilities are in a period of transition. The demand for electricity has been growing only about half as fast as it grew before 1973. Moreover, the costs of providing electricity are increasing rapidly. Nuclear power has not proved as economical, trouble-free, or popular as had been expected. New strategies are clearly needed to utilize capacity efficiently, develop new facilities at low cost, and to introduce pricing policies that will advance these goals. Questions even have been raised about introducing competition. There appears little likelihood, however, of any major changes in the near future in the structure of the industry, such as complete deregulation or splitting the industry into the three main components: generation, transmission, and distribution.

Policy Alternatives

The United States has developed three types of tools to deal with problems of production and distribution in the economy, and all have been applied at some time in energy markets.

Where competition can work—but where it may be threatened by the power of aggressive firms or by collusive agreements—the government has sought to preserve competitive conditions by enforcing antitrust policies.

Where competition cannot work because a natural monopoly exists—a characteristic of most electric and gas utility distribution systems—government has sought to regulate the price at which the service is provided to protect consumers and to promote the efficient operation of monopolistic enterprises.

Where incentives are not strong enough to induce private firms to provide certain essential services or where there are other public policy considerations, government has made the case for public enterprises, such as hydropower projects, or has subsidized research and development for nuclear power and, more recently on a modest scale, solar energy.

Sometimes, too, government has sought to intervene in ways that were later widely criticized, such as the imposition of oil import quotas and price controls on natural gas production.

What is clear in principle regarding the role of government may be very ambiguous in complex cases. To regulate or deregulate is often a debatable issue. And the choice that wins the majority support in one period may be abandoned when circumstances change. Technological changes, for example, may undermine the rationale for regulation by removing conditions that had previously made competition unworkable. Short-term interruptions in supply may engender demands for government intervention to help suppliers or consumer groups in ways that may inhibit desirable long-run adjustments, and firms, responding to immediate pressures or in anticipation of changes, may take defensive or aggressive steps that may have unanticipated deleterious effects on competition.

It is probably fair to say that neither antitrust, nor regulation, nor public enterprise has worked as well as its adherents had hoped. It is also true that a vast literature on the subject reveals persistent differences of opinion about the efficacy of different approaches. A prevailing consensus favors competitive markets, but the variety and complexity of energy markets suggest a continuing need for the examination of alternative public policies and the more effective implementation of them.

BIBLIOGRAPHICAL NOTE

A readable and instructive, though dated, introduction to the analysis of competitive conditions in American industry appears in Walter Adams, ed., *The Structure of American Industry: Some Case Studies* (rev. ed., New York, Macmillan, 1954). See especially the chapters by Joel Dirlam on the petroleum industry and Jacob Schmookler on the bituminous coal industry.The concepts used for analyzing competitive conditions for public policy purposes are explained by Richard Caves in *American Industry: Structure, Conduct, and Performance* (4th ed., Englewood Cliffs, N.J., Prentice-Hall, 1977). The analysis of the effects of vertical integration on profits in the petroleum industry appears in Harry G. Broadman, "Intraindustry Structure, Integration Strategies, and Petroleum Firm Performance," (Ph.D thesis, University of Michigan).

Promoting Competition in Regulation Markets, (Washington, D.C., The Brookings Institution, 1975), edited by Almarin Phillips, contains two chapters that challenge some prevailing views about regulation of utilities, one by Walter J. Primeau, the other by Leonard W. Weiss. Two chapters in *Case Studies in Regulation: Revolution and Reform,* edited by Leonard W. Weiss and Michael W. Klass (Boston, Little, Brown, 1981) are helpful on recent developments—one by Ronald R. Braeutegam on natural gas, the other by W. David Montgomery on oil. For a broad and systematic textbook treatment of many of the issues discussed here, see William G. Shepherd and Clair Wilcox, *Public Policies Toward Business* (6th ed., Homewood, Ill., Richard D. Irwin, 1979).

—— CHAPTER 6 ——

Energy
and the Environment

Energy consumption and environmental pollution have increased hand-in-hand. As nations became industrialized, they consumed ever increasing quantities of fossil fuels, thereby adding to air and water pollution, despoiling the land, and posing safety and health hazards for workers on the job. The pattern is familiar and it is worldwide. But only in the past two decades has it been recognized as a major public policy problem.

Not that all environmental pollution is attributable to the use of energy. Mining and manufacturing processes of all kinds generate pollutants, and so do consumers in their daily living. Workers outside energy industries, as well as those within it, face serious hazards on the job, particularly those employed in construction and the manufacture of toxic chemicals. Nevertheless, the process of extracting fossil fuels such as coal and oil, transporting them, and burning them is an identifiable significant cause of today's environmental difficulties, especially air pollution.

The major alternatives to fossil fuels—nuclear energy and hydropower—also pose environmental threats, some of them of quite a different nature than those posed by fossil fuels. All fuels considered, environmental costs are associated with virtually all the energy used in the United States.

Does this historical experience indicate that economic growth, energy consumption, and pollution are linked in some unalterable fashion—that if the GNP is to increase by 2 percent, energy use and pollution will have to rise at the same rate? When the seriousness of environmental problems became recognized in the 1960s, many people seemed to accept that view, arguing that environmental quality could be improved only by reducing the use of fossil fuels and cutting economic growth. But it has become increasingly apparent that there are other possibilities; that economic and environmental objectives may be pursued simultaneously—at least up to some point.

Over the years it has been possible to get more work done per barrel of oil and ton of coal, as shown in Chapter 2. After 1975, for example, rising prices and other pressures led to greatly increased efficiency in energy use.

Similarly, there is no fixed relation between energy consumption and the amount of pollution that is produced. The environmental impacts of energy are affected by the choice of fuel, modifications in industrial processes, and installation of pollution control equipment.

Thus it is a mistake to assume that we must choose between a cleaner environment and a growing economy. Our options are more numerous and our choices more complicated, as this chapter will show. Essentially, the problem for public policy is how to obtain and use the energy we need for economic growth without sacrificing our environmental goals. Not an easy balancing of objectives but it is not an impossible one.

POLLUTION IN PERSPECTIVE

Three aspects of the energy–environmental relationship need to be made clear at the outset.

First, it is neither feasible nor necessary to eliminate all sources of pollution. The ecosystem is capable of absorbing a substantial amount of the by-products of combustion that are discharged into the air and water—indeed, it has done so throughout history. Pollution has become a serious global problem in recent decades because rapid industrialization has led to an increase in the rate of pollution that strains the absorptive capacity of nature, and because certain technologies have produced goods and wastes—some of them highly toxic—that cannot be recycled by natural processes.

Second, the size and seriousness of the threat to the environment is difficult to assess with available information. We can actually see some of the environmental degradation—eroded hillsides, dirty air, dirty water, and dying lakes. We can smell or feel some of the pollution. But many pollutants—probably the most serious ones—are neither visible nor detectable by our unaided senses. They can be measured only by very sophisticated scientific tests, and their effects on the environment and people may not become apparent for many years. Furthermore, the system for measuring and assessing changes in environmental quality is not comprehensive enough, and there is a lack of reliable data about the nature of environmental hazards—from the by-products of combustion to toxic chemicals—and about air and water quality throughout the country, which is a great obstacle to environmental policymaking.

We are even more in the dark regarding the link between pollution, on the one hand, and damage to the environment or to human health on the other. Thus, although there is little doubt that the environment has been degraded and that such degradation poses a threat to human health, to plant and animal life, to land, and to water, we frequently lack the information we need to assess the seriousness of the threats, to devise countermeasures, and to decide how much effort should be made to maintain or restore various environmental values.

Third, an economic perspective tells us something important about the reasons for and the cost of pollution. To a large extent, environmental degradation has increased because polluting is cheaper for the polluter than other methods of waste disposal. The choice between dumping wastes into the water, or air, or onto vacant land and spending money for recycling, cleaner fuel, or modifying the production process was usually resolved in favor of dumping.

The economist states the proposition more formally. Air and water—and some lands—are "common property resources." That is, they are not owned by individuals or organizations, and the services they provide are widely available without charge. One of these free services is providing a dump for the by-products of human activity. Fuel-burning equipment—in homes, autos, industrial plants, and utilities—discharges some pollutants into the air. At no cost to the fuel-user, the air carries away these gases and particles. Similarly, discharging wastes into rivers and lakes is costless to water polluters. If the wastes are too great to be cleaned up by natural processes, pollution builds up in the air or in the water. But this may be of little concern to the polluter since costs of the discharge are borne by others downwind or downstream, or by the general public, who must either accept the pollution or spend to clean it up.

The problems of common property resources are not limited to pollution, of course. A classic example of the problem is that of international fisheries. Since no one owns international waters, each nation that fishes the area has an incentive to maximize its catch. The more that one nation catches, the better off it will be in the short run, but reducing one's catch may not improve one's own prospects for the long run if other nations pull out as many fish as they can. There is no incentive for any country, acting alone, to conserve—to limit catches to ensure that there are fish in the area in the future. The end result is a steady depletion of fishing resources unless an international agreement is reached to control the catch. The threat to the global climate from an increase in coal consumption, discussed later in this chapter, poses a similar problem.

Management of common property resources, in short, calls for government intervention because these resources are jointly owned and cannot be divided into units that can be bought and sold in the market. We must decide collectively what we want to achieve and by what means we wish to carry out our policies. The means, as will be shown later, may include devices that will make marketlike exchanges possible.

Pollution can also be described as an unpaid cost. In principle, the full costs of manufacturing a product include not only the capital and labor needed, as well as the materials required, but also the disposal of the wastes that are left over from the process. If air and water were not freely available, the manufacturer—and, eventually, probably the consumer—would have to pay the cost of disposing of wastes. However, access to common property to use as a sink enables the polluter to escape the full cleanup costs. He does not need to include them in the price of the product.

The selling price of a product which does not include disposal costs will be lower than it would otherwise be and, as a result, consumption of the item is likely to be greater. For some products, such an increase in consumption may not be important, but with regard to energy, it is important since energy policy in recent years has sought to reduce consumption and thereby reduce dependence on foreign sources of supply. If energy prices are artificially low, they are likely to encourage waste rather than conservation. As noted elsewhere in this book, however, neither lower energy consumption nor higher energy prices are in themselves desirable policy goals. They become desirable only as they reinforce other policy decisions.

It is worth noting parenthetically that on this issue environmental policy and energy policy are complementary. To the extent that environmental cleanup can be charged to energy producers, the price of energy will rise. A rise in price to cover costs will tend to encourage conservation. Lower consumption, in turn, will further reduce the discharge of pollutants.

With these preliminary considerations in mind, we are ready to explore the environmental consequences of particular types of energy consumption and the governmental steps that have been taken to cope with them.

ECOLOGICAL AND HUMAN COSTS OF ENERGY

Every energy source has its distinctive set of side effects. Some effects of energy consumption on the environment are immediate and transitory, others are persistent and cumulative. Some threaten human life and health directly, others only indirectly by damaging flora and fauna or by interfering with natural ecological processes. Some side effects are easily prevented or ameliorated, while others are rather resistant to human intervention. Our understanding of the nature and seriousness of these effects influences not only our choices of what fuels to use for what purposes, but also our priorities for research and development and our attitudes toward government regulatory policies.

Coal: The Price of Plenty

Coal, our most abundant source of energy, is also the fuel that causes the greatest array of environmental problems. Underground mining threatens miners' safety and health and leaves the land vulnerable to cave-ins; strip mining defaces the land and exposes it to erosion; and drainage from abandoned mines pollutes rivers and streams with acids. Burning the coal to obtain energy releases pollutants into the air that are harmful to human health; some of these pollutants may be carried considerable distances by the wind before settling to earth, often in the form of acid rain that pollutes rivers and streams, thereby endangering fish and plant life. In the long run, an increase in the combustion of coal and other fossil fuels may alter the climate of the earth.

To point out that the use of coal has this wide range of potentially damaging effects is not to say that intolerable damage will occur. Indeed, many ameliorating steps have been taken to reduce pollution and occupational hazards. Rather it is to indicate the breadth of the problem and the magnitude of the dificulties that need to be overcome if an increasing amount of energy is to be obtained from coal.

What worries scientists the most about an increase in the burning of coal is the possibility that the global climate may become much warmer because of "the greenhouse effect"—the buildup of carbon dioxide in the air, which blocks the escape of heat from the earth into space. Shortwave radiation from the sun can continue to pass through the carbon dioxide to the earth, but much of the infrared heat from the earth can no longer escape.

Until the middle of this century, increases in the atmospheric concentration of carbon dioxide were too small to cause concern. But during the past two or three decades the concentration of carbon dioxide in the atmosphere has increased measurably (by 7 percent over the 1958–79 period, according to the Council on Environmental Quality's annual report for 1980), arousing considerable concern about the greenhouse effect and its implications for world climate.

Scientific studies suggest the possibility that over time—perhaps by the middle of the twenty-first century, depending on the rate of fossil fuel consumption—the average temperature of the earth's surface may be raised by 3° or 4° Celsius, considerably more at the poles. If this change takes place, wind and rainfall patterns and ocean currents could be altered, with a pronounced effect on agriculture and on ecological systems in general. Such a change could perhaps prove beneficial for some regions, but disastrous for others. A matter of special concern is the fact that climatic changes brought about by the accumulation of carbon dioxide are likely to be irreversible.

Much remains to be learned about the relationship between fossil fuel combustion and climate change, but sufficient evidence has been accumulated to lead prudent observers to urge more aggressive monitoring of carbon dioxide and further research on its atmospheric effects. A report of the National

Academy of Sciences in 1981 warned that the major consumers of fossil fuels must begin planning now to develop alternative sources of energy so that they will be prepared to shift to other fuels as evidence of incipient climate changes indicates the need for policy responses.

Coal combustion also contributes to another type of pollution—acid rain—that is damaging the ecosystem in the eastern United States and Canada, and parts of Western Europe. Carried northward and east by prevailing winds, sulfur and nitrogen oxides from the combustion process combine with the moisture in the air to fall as acid rain, which happens to be landing on some of the choicest outdoor recreation areas on the continent. Lakes and topsoil in the Northeast are especially vulnerable to dry acidic deposits, as well as acid rain, because they are not underlain with limestone and other alkaline rocks that would help buffer the acid. It has been estimated that more than one hundred lakes in the Adirondacks area are dead—no longer supporting living organisms—and that in Nova Scotia nine rivers have become so acid that salmon can no longer reproduce in them. In many areas several important species of fish and other marine life have been eliminated, and substantial damage to crop plants has been reported.

Precise measures of the scope and seriousness of the threat are not available, but current evidence indicates that particular regions rather than all areas of the world are affected since the oceans cover most of the earth's surface and are sufficiently alkaline to absorb large quantities of sulfur and nitrogen oxides without damage. The weight of scientific opinion agrees in general on the origins and consequences of the acid rain phenomenon, but there is considerable disagreement on particulars and on the seriousness of the damage. The need for further study of the phenomenon and its consequences is one issue on which there is little disagreement.

There are some riddles. For example, the prevalence of acid rain and the awareness of its damage has occurred during the past five to ten years even though discharges of sulfur oxides, considered the principal culprit, have not increased. This, on the face of it, is puzzling. A National Academy of Sciences report has offered some hypotheses that may resolve the paradox—and that also illustrate how well intentioned efforts to control pollution may have undesirable side effects. First, because of the installation of high stacks to reduce pollution in the immediate vicinity of a coal-burning power plant or copper smelter, gases are injected into the air at a higher altitude than in the past; they remain in the atmosphere longer and are carried farther. Second, regulatory requirements, aimed at reducing the discharge of particulate matter, have led to the installation of equipment that filtered out one pollutant, alkaline fly ash, that in the past may have partially neutralized the sulfur emissions. And third, the rise in use of air conditioning has increased the demand for electricity in the summertime when high temperature and humidity may speed conversion of gases to sulfuric acid. (Other changes in weather conditions also affect the damage from acid precipitation. Melting snow, for example, spreads concentrated acidic accumulations in the spring when fish and plant life are young and especially vulnerable.)

Acid rain is an especially sensitive and controversial issue because those who bear the brunt of the damage are often not the ones who benefit from the mining and burning of coal. The beneficiaries of coal production—mining companies and workers in Illinois, Indiana, Ohio, and Pennsylvania principally—have fought certain environmental regulations, fearing that higher air quality standards would persuade utilities and others to switch from high-sulfur coal that they produced to low-sulfur western coal—a shift that could destroy jobs and bankrupt businesses in the Midwest. Meanwhile the losers downwind, especially the Canadian government, have called for the imposition of tighter controls. One industry response has been that since only a small proportion of the discharges from coal-burning plants is implicated in acid rain, it may be more cost-effective to combat acid rain where it occurs through liming of lakes rather than at the source.

The more familiar consequences of coal combustion on air quality are its direct effects on visibility, human comfort and health, and property. While available data are not adequate to indicate precisely the proportion of air pollution attributable to coal, enough monitoring has been carried out to identify it as a major contributor. Coal is a major source of sulfur dioxide, which may aggravate respiratory ailments. In the atmosphere, it may be changed to sulfates, which are more hazardous to health. Coal-burning plants also emit particulates—small particles of minerals, some of which are highly toxic when inhaled. At what concentrations these pollutants interfere with breathing, cause illness, or even death is not known, but the lethal effects of acute episodes have been confirmed. At Donora, Pennsylvania, where sulfur dioxide and particulates reached unusually high levels in 1948, illness rates were put at 10 percent and the death rate at one-tenth of 1 percent of the exposed population. Less well understood are the effects of exposure to lower levels of pollution over time. Coal is the principal fuel for power plants, which are responsible for about half of the nitrogen oxides in the air. Nitrogen oxides have also been linked to respiratory illness, as well as to the formation of smog, and acid rain. Pollutants from coal combustion also add to the soiling of clothing and other property.

Coal mine safety is an area where considerable progress has been made. From 1970 to 1979—the decade after enactment of the Coal Mine Safety and Health Act of 1969 (amended in 1977)—the number of fatalities was cut in half. The decline in the death rate was even steeper. For each million hours worked, the death rate in 1980 was only one-third of what it had been in 1970. The change in the number of fatalities over a half-century was more dramatic. There were 105 fatalities in 1978 compared with more than 1,000 in 1931. Injuries in 1978 totaled 13,554, compared with nearly 100,000 in 1931—an enormous improvement even after taking into account the growing importance of strip-mining.

Despite stricter federal standards for ventilating explosive mine gas and coal dust and for preventing collapse of roofs, underground coal mining is still one of the nation's most hazardous occupations, as the nation was reminded during 1981 and 1982 when the accident rate turned up.

That safety in underground mining can be improved further is suggested by injury rates that vary enormously from mine to mine. For the safest mines, accident rates are about the same as for retail trade and education, but for the least safe, they are tenfold higher. Improvements in mine safety pay off not only directly in lives saved, but also in the reduction of suffering in miners' families and in the costs of compensation paid by society to the disabled and to surviving wives and children.

The principal health hazard, black lung disease (pneumoconiosis), has exacted a heavy toll for decades. According to a Bureau of Mines study in the 1970s, it would affect 15 percent of miners then living. Efforts have been made to reduce the incidence of this disease by reducing coal dust and to compensate the victims, but it remains a major concern for all who work in the mines.

Oil: The Cost of Moving and Using It

The blowout of an oil well off Santa Barbara, California, in 1969 drew attention to the environmental costs of exploiting offshore oil and was one of the events that generated support for Earth Day in 1970. Similarly, the sinking of huge tankers in recent years has dramatized the vulnerability of the ocean to pollution from petroleum. Less dramatic but far more damaging to the environment are routine emissions from autos and from industrial plants that burn oil. Though oil is a more benign fuel than coal, it is consumed in far greater quantity than any other source of energy and is the principal fuel for our transport system.

Costs of controlling the Santa Barbara blowout—cleaning up the beaches, plus damage to the fishing industry and recreational areas—were comparatively modest (being estimated at $16.5 million), compared to the emotional shock effect of the despoiling of a lovely beach and the killing of seabirds, including half the loons and grebes in the Santa Barbara channel. Similarly, the economic cost of a much larger spill later in 1969 off the coast of West Falmouth, Massachusetts, was dwarfed by the ecological effects. Killing of plant life and birds continued for eight months, wiping out half of the area's living matter. There have been other blowouts of larger and smaller dimensions—sometimes accompanied by fire—off the California coast and in the Gulf of Mexico. The worst was the blowout of a well owned by Pemex, the Mexican National Oil Company in Mexico's Bay of Campeche in 1979—a year in which the amount of oil spilled was three times the amount spilled in any previous year.

Since 60 percent of the annual world output of oil moves across the oceans, tankers in transit pose a worldwide spill hazard, as such spectacular episodes as the spill from the *Amoco-Cadiz* in the English Channel in March 1978 have reminded us. But large accidents are responsible for only a small proportion of the total oil released in the ocean. The ordinary process of discharging ballast from tankers has been a more serious cause of pollution. After ships unload their oil cargoes at refineries they are customarily filled with water for the return trip. Since unloading is never 100 percent, some oil remains in the tanker and mixes with the water. This oil is partially skimmed off before the water is emptied out after the return trip, but some of the oil remains and is discharged into the ocean. Even though the individual discharges of oil are small, collectively they are significant because so many huge tankers have to be deballasted. Improved practices are being introduced, however, to reduce pollution from deballasting.

The full, long-range ecological damage of blowouts, tanker sinkings, and spills has been difficult to measure. Scientific studies suggest that the recuperative capacity of the oceans after a single large spill may be greater than it was thought to be years ago. For example, aquatic life was virtually back to normal five years after the Santa Barbara spill. Meanwhile, it has become apparent that chronic, small spills may be more damaging because they seem to prevent the natural system from ever recovering fully.

By and large, the oil companies have borne the major direct financial costs of these accidents—losses of oil, costs of cleanup, costs of control, and so on—but costs have been imposed on the environment and on the rest of society as well, and some of the cleanup costs may have been passed along to buyers. Proposals for a surcharge on shipments of oil or, in the case of offshore wells, on extractions of oil have been put forth as a way to internalize the costs of these accidents—that is, to build them into the price of oil products—but so far none have been enacted.

Environmental damage from producing and shipping oil, though substantial, is much less than damage from burning it. Transportation, particularly by auto and truck, is a major source of air pollution. Autos and trucks account for more than half of our annual consumption of petroleum and for most of the carbon monoxide that is discharged into the atmosphere from all sources, as well as for much of the hydrocarbons, nitrogen oxides, and particulates.

If combustion were chemically perfect in an automobile, there would be no discharge of carbon monoxide; but the internal combustion engine is relatively inefficient, and it has been estimated that the discharge of carbon monoxide is at the rate of a half-ton to one ton per driver year (four times the discharge from a diesel engine). When a car is idling, its discharge of carbon monoxide more than doubles. Thus, in crowded streets pedestrians and passengers can be exposed to a high level of carbon monoxide. Faulty mufflers or leaks may also increase the exposure of car drivers and passengers. But although carbon monoxide is harmful—it is absorbed unusually rapidly by the blood and can affect awareness and response rates—it probably does not pose

Environmental Hazards Indoors

Energy use affects air quality in residences and public buildings as well as outdoors, but so far no systematic efforts have been made to manage the "indoor environment," except in the workplace. Indeed, little attention was focused on this issue until the National Academy of Sciences, at the request of the EPA, undertook a study of the nature of the indoor environment problem in 1979. The study sought to monitor gases emitted from building materials and furnishings, substances such as asbestos, and the products of combustion from gas stoves, wood fires, and tobacco.

The NAS report cites studies showing an association between gas cooking and impairment of lung functions in children. It also showed that in busy urban areas, exhaust from autos raises pollution levels in residences and offices. Other data show that the concentration of carbon monoxide and nitrogen oxides indoors may exceed the outdoor ambient air quality standards.

Problems of indoor pollution have been exacerbated in some buildings by the steps taken in recent years to conserve energy by increasing insulation, sealing windows and doors, failure to vent certain discharges, and generally reducing the interchange of indoor and outdoor air.

The NAS study concludes that pollution levels in homes and offices "may constitute sufficient threat to general public health to justify remedial action." Further study and monitoring are also needed for enclosed areas open to the general public—museums, transport terminals, theaters and arenas, schools, and buses and trains.

a long-run health or environmental threat. Recovery is quick out of doors; with a reasonable breeze, fumes are rapidly dispersed. Carbon monoxide from an auto is likely to be fatal only to someone in a closed garage with the engine running.

Auto emissions are also the source of half the nitrogen oxides in the air and a major source of hydrocarbons and particulates. Thus, together, cars and power plants account for most of our air pollution. In the near future, the emission of particulates from automobiles is likely to become an increasing hazard. The major U.S. auto producer, General Motors, has announced plans to increase its production of diesel-burning vehicles, which emit not only more particulates than comparable gasoline-burning cars but also a higher proportion of small particulates that are more likely to penetrate the lungs.

Determining what is a tolerable level of emission for the average vehicle is difficult because it is not just the emission level that determines ambient air quality. The concentration of vehicles and the opportunity for dispersal of pollutants is also important, as borne out by the experience of Los Angeles which has suffered from smog brought on by unusually heavy auto use and the action of the sun on the pollutants often trapped for long periods over the city.

Burning oil to heat homes and run power plants, like burning coal, also contributes heavily to air pollution by producing nitrogen oxides, sulfur oxides, hydrocarbons, and particulates.

Nuclear Power: The Generation of Controversy

Nuclear reactors produce none of the familiar pollutants that have made fossil fuels such a threat to climate or clean air. The nuclear hazard is radiation—in small amounts from uranium waste (tailings) at the mine and mill, from the operation of nuclear reactors, or from shipment and storage of waste—and in potentially large amounts from reactor accidents.

Exposure to radiation can lead to cancer, to genetic mutations, and in extreme cases, to death. This much is certain. But the links between the size of a dose, frequency of exposure, and elapsed time between exposure and effects are poorly understood, and they have been the subject of bitter controversy. Thus, even though high doses of radiation are clearly hazardous, the amount of low-level radiation that can be absorbed without damage is not known. Cosmic rays from outer space and radiation from the earth establish a level of background radiation to which we are all exposed. Modern technology, such as medical X rays and television, increase the exposure to low-level radiation. Radiation from routine reactor operations appears to add very little to the total exposure of the U.S. population, or to pose a health hazard for persons living near nuclear plants.

Other hazards of a more familiar type are also associated with nuclear power. Reactors require water for the cooling phase of the power cycle—about twice as much water as comparable coal-burning plants—and unless the water is permitted to cool again before it is discharged it will raise the temperature of rivers and lakes, with deleterious effects on some plant and animal life. Waste from uranium mining, like wastes from coal mining, can pollute surface water, though the nature of the pollution—radioactive substances—differs from the acid drainage associated with coal. Like coal miners, uranium miners face an occupational hazard, in their case, from the exposure to radiation.

During the first three decades of the commercial use of nuclear power, the known adverse effects of nuclear energy on people and on the environment have been dramatically less than those from fossil fuels. Radiation from operations has been reasonably well controlled, and, after years of neglect, so has radiation from mining. The distinctive characteristic of the nuclear hazard, however, is its *potential* for harm—not what has happened, but the possible catastrophic dimension of what might happen. This is what makes comparisons between fossil fuels and nuclear power so difficult. Possibilities of enormous damage and unknown consequences must be weighed against the identifiable, continuing, and familiar damage from the use of coal, oil, and hydroelectric dams.

The accident at the Three Mile Island nuclear plant on March 28, 1979, provided an especially instructive insight into the nature of the risk associated with the operation of nuclear power plants. As various studies have pointed out, the accident began with a minor mechanical failure of a routine type that was correctable. It worsened because of operator failures, instrumentation problems, and managerial mistakes. That such a malfunction could occur was a recognized possibility, but the sequence of human responses to the accident was not predictable, and the dimensions of the episode itself were surprising. Previously it had been widely assumed that if there were a large loss of coolant in a reactor, either the process would be stopped *before* the reactor core was damaged or there would be a meltdown that would allow radioactive material to escape. An accident in which there was substantial core damage but only moderate meltdown had not been considered a likely outcome. But it apparently occurred.

Among the lessons of Three Mile Island—and volumes are filled with analyses and suggestions that should contribute to greater nuclear safety in the future—an important one is that there is still a lot to learn about the range of possibilities and the nature of nuclear accidents. Three Mile Island was full of surprises, and more surprises can be expected from future incidents. It served as a reminder that machines and people interact, and that the training, judgment, and alertness of plant operators is highly important. It also left the debate unsettled. Pronuclear forces pointed to the absence of severe damage in the aftermath of a massive failure, while antinuclear advocates stressed the potential for disastrous accidents and their unpredictability.

Apart from the possibility of a disaster, minor accidents and problems of various types that have been reported periodically in the press have served to perpetuate public uneasiness about nuclear power. At the Diablo Canyon plant of Pacific Gas and Electric in California, the start-up was delayed in the early 1980s by the discovery of a design error that had left the plant vulnerable to an earthquake. At Ontario, New York, a valve that had been opened to combat a leak caused by a burst pipe stuck open, causing a temporary threat to the reactor core. In addition, "thermal shock"—the possibility that a reactor vessel might crack because of a sudden change in temperature—was identified as a likely hazard at some plants because of evidence that the welds on a vessel have a shorter lifetime than originally believed.

Even before Three Mile Island the rosy future once seen for nuclear power had become clouded. As discussed elsewhere in this book, the growth of demand for electrical power had slackened, costs of construction had risen, the fear of proliferation of nuclear weapons made possible by the presence of power reactors had become an important political issue, and the delay in establishing a plan for permanent disposal of nuclear wastes had led to widespread public concern. Of these issues, the one closely linked to environmental concerns is waste disposal.

Although the need for "permanent" storage of nuclear waste has long been recognized, nuclear waste continues to be stored above ground in temporary facilities of varying reliability. ("Temporary" has come to denote a period of several decades!) Utility power plants store their spent fuel in adjacent water pools. At defense plants, some waste used to be stored in tanks, and leaks were reported as far back as the mid-1960s when there were leaks at the Hanford, Washington, installation. A good deal of treating and repackaging has since reduced but not eliminated the problem.

From time to time proposals for permanent disposal have been set forth. They have incorporated plans for storage in leakproof containers with burial in deep man-made caverns in geologically stable formations, such as salt mines or perhaps granite formations. Burial is still considered to be the most likely long-term solution, but debate continues over such matters as the type of containers, the best geologic formations, and the kind and extent of damage in the event of an escape of radioactivity. Another issue is whether to bury either (a) the spent fuel rods containing fissionable material that could be reprocessed and reused in reactors, or (b) the possibly less radioactive residue left over after reprocessing.

One major study concluded that because of uncertainties, the decision about ultimate disposal should be postponed until the 1990s. Meanwhile, the industry will continue to rely on storage above ground, even though warnings have been sounded that some storage pools may be filled before alternative disposal sites are available (which could force some plants to shut down) and despite the controversies aroused by trucking wastes from plant sites to other available temporary storage sites.

Hydropower and Other Gentler
Alternatives

Generating electricity by harnessing the power of flowing water seems harmless enough. Neither combustion nor toxic discharges, are associated with hydropower, nor are occupational diseases. Nevertheless, though cleanliness and dependability are indeed attractive features, hydropower is not entirely benign. It does entail environmental costs. They are not of the same sort as the damage from fossil fuels, but from an ecological perspective they can be more serious.

The building of huge dams for electric power has led to landslides in some countries, earthquakes, a rare breach or collapse of dams, and waste of water in areas where it is vitally needed, as well as to undesirable biological consequences. In the western United States, for example, there has been substantial loss of water because evaporation from reservoirs created by hydropower projects is greater than from flowing rivers. According to one estimate, the loss from evaporation on Colorado River projects alone could supply the water needs of 4 million people. Another consequence of evaporation is an increase in a river's salinity, which reduces its suitability for drinking and farm use.

Changes in streamflow and water temperature may affect the balance among different species of fish or threaten the survival of some—a matter that can arouse a great public outcry, as demonstrated by the battle to save the snail darter's habitat from destruction by the Tellico dam project in Tennessee. Proliferation of weeds in reservoirs may clog inlets and outlets and spoil fishing and recreation. Even where boating and swimming facilities are increased there may be an aesthetic price, the loss of a "wild" river.

The most benign potential source of energy is the sun, though not all forms of solar energy are equally gentle on the environment (nor economically feasible). Large centralized solar technologies are less desirable ecologically than decentralized solar systems, such as roof collectors for home heating and water heating. In dry regions, which are especially attractive for large solar development, the land surface is fragile. Breaking the top crust will make the land more vulnerable to wind erosion. Other consequences that have been identified are much less important than those linked to the use of fossil fuels: paving and construction may interfere with the growth of vegetation and destroy the habitats of burrowing animals; glare from solar collectors may blind or confuse birds; combustion of biomass, like the burning of fossil fuels, may add to air pollution; construction of ocean thermal plants may affect ocean temperatures and currents, and, indirectly, marine life. Thus the installation of some solar energy devices will call for precautions, but there is no question about the natural superiority of solar power from an environmental perspective.

ISSUES AND POLICY ALTERNATIVES

During the so-called environmental decade of the 1970s the United States enacted wide-ranging legislation to reduce air and water pollution and to control other environmental hazards. New agencies such as the U.S. Environmental Protection Agency (EPA) and the Occupational Safety and Health Administration (OSHA) were set up to enforce the new legislation. State and local environmental agencies were also established. Environmental interest organizations became increasingly visible participants in policymaking and in enforcement, especially through court suits challenging private and public decisions and the failure of government agencies to carry out legislative mandates. By the 1980s, energy-related activities were affected by regulations across the board—from extraction of fossil fuels to disposal of wastes. Some regulations resulted from legislation directly aimed at energy-related activities; and some from provisions of broad environmental laws aimed at protecting a resource, such as water.

Federal Laws and Regulations

The list of legislative enactments is impressive in its scope. For example, the Coal Mine Health and Safety Act of 1969 imposed regulations that would protect workers from mine hazards. The Surface Mining Control and Reclamation Act of 1977 sought to protect the land from erosion and other side effects of strip mining and underground mining, such as contamination of water. The Resource Conservation and Recovery Act of 1976 regulated waste disposal and encouraged experiments to extract fuels from garbage and other solid waste. The Marine Protection Research and Sanctuaries Act of 1972 (the Ocean Dumping Act) prevented the United States from disposing of nuclear wastes and other "harmful" materials into the ocean and limited the dumping of other wastes as well.

More far-reaching in scope were the 1972 Amendments to the Federal Water Pollution Control Act which were aimed primarily at controlling sewage and industrial wastes, including wastes from petroleum refineries and thermal pollution from power plants. Since stricter environmental controls tended to raise the cost of energy, the amendments even included a provision requiring that energy efficiency be taken into account as one of the criteria for awarding grants to construct sewage treatment plants.

The water control amendments were concerned not only with pollution of surface water but also with groundwater—the water below the surface in permeable rock formations, which supplies about one-fourth of the nation's needs for fresh water. Groundwater pollution from petroleum and coal mining has been documented in seventeen states, though it is relatively small compared to other sources of contamination, particularly from synthetic organic chemicals. But surface water pollution by energy-related activities has been more serious.

Under the Clean Water Act of 1977, further amendments were enacted to improve the effectiveness of the water pollution control programs, though they did not greatly alter the long-run objectives nor regulatory approaches previously established. However, little was done to regulate so-called nonpoint sources of pollution—that is, pollution of lakes, rivers, and streams by the runoff of waters from city streets during storms and by the runoff of topsoil and chemicals from commercial fertilizers used on agricultural lands.

The legislation with the most far-reaching consequences for energy use was the series of clean air acts passed over a period of two decades. The Air Pollution Control Act of 1955, which established federal interest in research on air pollution, had emerged from the reaction to two dramatic events—the Donora episode and the buildup of photochemical smog in Los Angeles, which scientists at the California Institute of Technology later linked to the effects of sunlight on auto emissions. After the London smog of 1952 and efforts by California to introduce emission control devices, sufficient support emerged in Congress for expanding the federal role in air quality regulation beyond research and training. The Clean Air Act was passed in 1963, and the Motor Vehicle Pollution Control Act was passed in 1965. The latter gave the Department of Health, Education and Welfare the power to set emission standards. Amendments in 1967 actually set national standards for automobile emissions and for discharges from major industrial plants. Responsibility for administering the clean air program was given to the states.

Disappointment with achievements under this spate of legislation led in 1970 to stronger amendments that identified a number of pollutants considered potentially hazardous to human health and specified that their concentration in the air should be kept at harmless levels. The 1970 legislation also introduced the controversial concept of "threshold value"—the notion that at some level the concentration of a pollutant would begin to affect human health; concentrations below the threshold level were assumed to have no harmful effects. The pollutants singled out for regulation included carbon monoxide, sulfur dioxide, nitrogen oxides, particulates, and oxidants—and later, lead. So-called primary standards (those required to protect human health) were to be set to provide a margin of safety below the threshold values. The act also required new industrial plants to meet stringent discharge requirements ("new source performance standards") regardless of the air quality in the area they were to occupy. Stricter standards were established for automobile emissions of hydrocarbons and carbon monoxide, and later for nitrogen oxides. States were to formulate plans to meet these standards.

What has been achieved by a decade of commitment to environmental regulation? There is probably a consensus that the deterioration of air and water quality and the degradation of land resources has been slowed, and that in some areas the environment has been improved. So-called dead lakes, like

Lake Erie, have been revived. In some areas the quality of air has been visibly improved. The EPA has reported a big drop in particulate emissions during the 1970s and a substantial decline in sulfur dioxide levels in the urban areas where it monitors air quality. And certainly environmental management has become an important concern for industry and government. No longer can it be said that pollution results largely from a lack of awareness of the environmental consequences of human activities or indifference to them.

However, some critics of the environmental control program have argued that at least some of the improvement is attributable to the decline in the use of energy in 1980 and the following years resulting from higher prices and the recession. Moreover, they contend that the costs of environmental improvement have been excessive because of the regulatory approach embodied in environmental legislation.

It is also true that on some matters little progress has been made. The battle against toxic substances, which may prove to be the most serious of our environmental hazards, has not advanced much despite the enactment of the Toxic Substances Control Act. Meanwhile, the emerging awareness of problems, such as acid rain and the greenhouse effect, have heightened concern about effects on the environment.

Whatever one concludes, the judgment is based to a large extent on fragmentary anecdotal evidence. There is no well-established data base to which one can turn for conclusive evidence about national or regional trends. The most readily available data are contained in the annual reports of the Council on Environmental Quality and the summary volume *Environmental Trends,* and these fall far short of what is needed. Special studies by the EPA and the National Academy of Sciences on air quality provide additional information, but often the results are highly tentative. Data about the emission of pollutants are limited and frequently based on engineering estimates. That is, instead of being measured directly, emissions are estimated on the basis of the amount of discharge that might be expected from certain kinds of equipment, using certain kinds of fuel, and operated for certain lengths of time. Air quality is often monitored at sites chosen for convenience rather than as part of a systematic program to obtain comparable measures in all urban areas; weather changes, time of day and seasonal factors have not been taken into account adequately. Repeated studies have pointed to the need for a much more effective system for data gathering and analysis. Not only are there gaps in data of all kinds, but, in addition, great uncertainty remains over a number of basic questions: Are the right (important) pollutants being measured? At what level of concentration do particular pollutants affect human comfort, health, and the death rate?

Another difficult question is whether the economic test is being met: Have we been getting our money's worth? Have realistic goals been set and efficient means adopted to meet them? Has the maximum environmental protection been obtained for each dollar spent for environmental improvement and how would one know? By the 1980s, these questions, long debated by environmental policy analysts, were becoming increasingly important in the political arena.

Questions of Costs and Risks

Early critiques of environmental policy pointed out three basic economic lessons that had been ignored by policymakers in formulating initial pollution control programs. These lessons are worth keeping in mind though some changes have been made in environmental policies.

First, it is relatively cheap to remove a substantial amount of a pollutant—say, 50 to 75 percent—but increasingly expensive to remove successive amounts. The tendency for cleanup costs to rise at an increasing rate has been documented in numerous industry studies. As one critique suggested, "Depending upon the industry or pollutant, going from, say, 97 percent to 99 percent removal may cost as much as the entire effort of going from zero to 97 percent." Yet early environmental legislation set such long-run goals for polluters as "zero discharge," which was not only unattainable technically but, if attainable or nearly so, would be exorbitantly expensive. Indeed, the question of cost was hardly taken into account in the early legislation. Standards of air and water quality were set based primarily on what was technologically feasible. (The "best practicable technology" or "best available technology" was to be employed.)

The second neglected economic lesson is that the cost of pollution abatement varies greatly from industry to industry, and even among firms in an industry. Thus regulations that require all industries and all firms to meet the same targets—which, offhand, may sound reasonable and equitable—turn out to be unnecessarily costly overall and impose excessive burdens on some industries. For example, suppose two firms or sewage plants happen to be responsible equally for virtually all of the pollution in an area, but that it costs one firm half has much as the other to reduce its discharges to an allowable level. More pollutants could be removed at less cost if the second firm paid the first one to carry out more than its share of pollution abatement. Such flexibility was not initially feasible, but is now being attempted by such strategies as the "bubble" concept discussed later in this chapter.

Third, there are many ways to reduce pollution. These include treating or removing pollutants at the end of the production process, making products out of different materials or by different methods, developing less harmful substitute products, and making it easier for the environment to absorb wastes by changing discharge points. The initial environmental legislation concentrated almost wholly on waste treatment, however, and this bias continues.

These lessons are closely related to the larger issue of how much can we afford to spend pursuing environmental goals. Cost considerations may be ignored for some hazards that society considers intolerable, but, in general,

environmental needs must compete with other needs for a share of the government budget, or of industrial expenditures. How much environmental protection we get depends not only on the amount we are willing to spend, but on
the cost-effectiveness of what is spent. The stakes are large. According to the
Council on Environmental Quality, annual expenditures for environmental
purposes are expected to increase during this decade from $37 billion in 1979
to $69 billion (in 1979 dollars) in 1988, a substantial portion of it to combat
the side effects of energy-related activities (see page 175).

Surveys over the years by private research groups and government agencies have led to widely varying estimates of this burden. But they all agree that
the costs of pollution control—especially air pollution—fall very heavily on
the energy industry and energy users. The electric utilities industry and petroleum refining together account for about half of all the capital expenditures
made for pollution abatement. For air pollution control alone, estimates show
that about 80 percent of total expenditures—capital expenditures and operating and maintenance costs—fall on the energy sector or are attributable to
energy consumption.

Along with questions of cost, environmental policymaking must reflect
a better understanding of the concept of risk. Early environmental aspirations
were strongly influenced by visions of a world free of air and water pollution
in which neither human health nor the ecosystem would be threatened by industrial activities and their by-products. Yet if we think about it for a few
moments, we realize that life continually confronts us with risks—some of
them rather commonplace, that decisions to accept or reject them have become
so much a habit that they are no longer made consciously, like crossing the
street. Human settlements, wildlife, and natural settings all over the globe are
exposed to natural disasters of some type—floods, drought, volcanic eruptions, earthquakes, diseases, and—decisions to stay put or move on depend
on balancing the risks against the advantages of staying on. Similarly, human
activities intended to achieve desirable goals entail risks. Although industrial
societies have learned to reduce or avoid some risks, they have introduced
others. The number and balance of risks that a society faces may change, but
a risk-free world is not attainable.

No satisfactory theory has yet been provided to explain why some risks
seem acceptable and others unacceptable, and why these assessments differ so
widely among people in a society. A number of hypotheses have been set forth.
For example, it has been suggested that people are willing to do to themselves
what they will not permit others to do to them. Cigarette smoking may be
hazardous, but it is a voluntary activity, and the risk is accepted by a large
proportion of the population. On the other hand, emissions from an industrial
plant, which may be a less serious health hazard but which are thrust upon us
may be intolerable, even to smokers. It has also been pointed out that people
react differently to hazards that are familiar than to those that are unknown.
One of the persistent anomalies cited in studies of natural disasters is that

people repeatedly driven out of their homes and communities by floods move back into the floodplain after the water recedes. Risk perceptions also offset decisions when tradeoffs—between jobs and safety, for example—are necessary.

It is not necessary to have a complete understanding of the ways in which people perceive risk and act on their perceptions in order to recognize the importance of taking the risk factor into account. What may at first appear to be a simple choice between pollution and no pollution is more appropriately viewed as a question, What is the risk to human health and to the ecosystem and how much, if anything, should be spent to reduce the risk, in comparison with the cost of reducing other risks? In short, environmental policy—as indeed all of our social and political policies—should ideally proceed from an assessment of how much risk is acceptable in a wide variety of circumstances. Such an exercise will not lead to a definitive conclusion, given the nature of the problem, but it can help to put the issue in a more realistic perspective.

Standards or Incentives

Direct environmental regulation, based on standards and enforced by administrative and legal means, has become enormously complicated and cumbersome. It places a heavy burden on government, which must conduct economic analyses of thousands of plants, study the technologies used in production, and assess the air quality in the area before it can impose controls source-by-source. It must then be able to defend in court its decisions regarding the choice of best available technology or other standard. This approach requires the government to become as expert on operating details as industry itself—not a very realistic notion. At the same time, direct regulation leaves the way open for industry to challenge specific decisions and delay the effective date of compliance. After carefully considering the odds against being caught, a firm may even choose the alternative of simply not complying. Since relatively few firms can be monitored for compliance, a polluter's risk of being caught may be slight. These are some of the major difficulties in enforcement that have been cited by critics of direct regulation.

The principal alternative to the present approach is to provide incentives for polluters to reduce their discharges and allow them full flexibility in going about it. Such an approach, according to its proponents, is less likely to invite legal bickering and more likely to encourage firms to invest in technology for environmental control.

One such system, which has been talked about for more than a decade in the United States and applied in modified forms in some other countries, is to charge emission fees. Energy users would be required to pay so many cents a pound, for example, for discharges of sulfur oxides, particulates, and so on. The fees would be based on the damage caused by the polluter (which, it must be admitted, is not easily determined). Firms could install pollution control equipment or change methods of production rather than pay the fees. By encouraging cleanup at those points where it can be done more cheaply, an effective system of emission fees or charges would make the total cleanup bill smaller for the whole economy. More of the job would be done by firms that can reduce pollution at a lower cost, less by firms which may find it cheaper to pay the charge rather than reduce the pollution.

Several objections to the emission fee, or pollution charges, have been made. One is summed up in the phrase "license to pollute." Some environmentalists have objected that the air and water will not be cleaned up if firms have the option of paying a fee to use them as dumps. But this ignores the reality that all control systems—including the direct regulation—are instruments for permitting some amount of pollution. Moreover, it ignores the incentive potential of a properly conceived fee system which would be designed to make pollution abatement more attractive than paying pollution charges.

Another criticism is that a fee system would require an extensive monitoring and enforcement apparatus, which is not presently in place. The point is well taken, but it applies with equal force to direct regulation. Much of the same information is needed when the government has to decide whether to grant a permit under the present system and to determine whether firms are in compliance.

One modest innovation that features an incentive approach was the EPA's adoption of the "bubble" concept, which permits a company to treat all the sources of air pollution in a plant as if they constituted a single source—that is, as if they were covered by a bubble. Within this bubble, increases in pollution by some sources can be offset by reductions in other sources so long as total output of pollutants by the firm does not increase. Thus management is given an incentive to reduce pollution where it is cheapest to do so and avoid expenditures for control equipment where costs are unduly high. This provides flexibility to management and promotes reductions in the total cost of meeting air quality goals.

Another innovation was the "offset" policy, which gives firms an opportunity to build new plants in areas that have not attained the air quality standards mandated by the Clean Air Act. Under the offset plan, a firm wishing to enter a "nonattainment" area can do so if it first installs control equipment and then can induce a firm already in the area to reduce its emissions sufficiently to compensate for the added pollution that would be caused by the new firm (or the expansion of a firm already in the area). The expanding firm is likely to get a taker on its offer if it can afford to pay more for the right than it costs the established firm to reduce its own pollution. This is a variant of the widely discussed incentive plan called "marketable permits" which seeks to arrange a market in pollution where one has not emerged out of the private sector.

Incentive systems can be designed to meet a number of problems of air quality regulation, such as regulation of new firms and the development of new technology. New firms are now required to install the "best available control technology" or the technology necessary to produce the "lowest achievable emissions rate." Such higher standards of pollution control make some sense, since new firms can generally incorporate pollution abatement equipment and procedures more cheaply than can old firms. However, to require that the higher standards be achieved by adoption of a single technology is misguided. It puts undue emphasis on only one way of reducing pollution. It forces widespread installation of equipment that may have a slight engineering advantage today but may be inferior to other processes in the long run. Moreover, emphasis on a technological criterion alone gives undue weight to small and perhaps insignificant differences in performance at the expense of other considerations. Suppose that the best available technology can remove 85 percent of a pollutant, for example, rather than the 80 percent that can be removed by another technology. It may be required even though the alternative technology may achieve its goal at a much lower cost. A fee system not only gives firms far greater flexibility, it also keeps the door open for the development of alternative technologies that will pay off in the long run. A fee system could also provide stiff penalties for malfunction.

The case for an alternative approach to pollution control rests not only on the potential advantages of an incentive system but also on flaws in the present system. For example, standards have been set for pollutants as if it were known that concentrations of a pollutant below the standards were "safe" for human health, whereas higher concentrations were hazardous. Yet it is clear from the continuing stream of studies about the relationship between air quality and human health that there is little scientific basis for these beliefs. The notion of a "threshold" has not been borne out by any convincing scientific evidence. The relative seriousness of various types of pollutants has not been established; whether those that have been singled out for control are the most threatening has been questioned. The link between exposure and illness or mortality has not been established, nor have the relative threats of brief high- level exposures and low-level exposures over prolonged periods.

Given such uncertainties, a regulatory program based on uniform compliance with absolute standards is hardly on firm footing.

Both direct regulation and a system of pollution charges are subject to limitations that come from the same source: the political environment in which the proposals are enacted. No legislation is likely to satisfy the specifications of the ideal prototype of either system. Constitutional and legal requirements will make compromises essential. Moreover a bureaucracy will have to be created either to oversee the monitoring of emissions and ambient quality, to establish allowable limits of discharges, and set fees, and to oversee compliance. Introducing a system of incentives would make possible considerable progress, but it will not be a panacea.

THE CLASH OF VIEWS

Energy and environmental objectives clash at many points, but this is not unusual in the public policy arena. There are conflicts between economic growth and price stability, between efficiency in the use of resources and equity in the treatment of different economic and social groups. There are conflicts between the sunbelt and the snowbelt, between the farm states and the urban centers. Thus the need to accommodate both environmental aspirations and energy needs is a variation on an old theme, not a new and uniquely troubling aspect of public policy.

The historical case is not much in doubt. Increasing industrialization and the adoption of auto transportation—both requiring greater energy consumption—have led to higher levels of air pollution and introduced or aggravated other environmental hazards as well. But the relation between energy and environmental goals in the future is much more fluid. Two factors will largely determine the extent to which both goals can be accommodated. One is technological progress—the extent to which fossil fuels can be burned more efficiently, emissions controlled more effectively and new, less polluting sources of supply put into place. The second is the trend of economic activity. A growing economy, though it may increase energy consumption, can afford to make expenditures for pollution control and development of new energy sources. One that is forced to choose between pollution and jobs is likely to put jobs first.

Both technological development and economic growth will depend in turn on the ways in which public policies on energy and environmental issues are directed. If rigid, technology-based standards continue to dominate environmental policy and if undue direct regulations govern energy policies, the likelihood of pursuing energy and environmental goals successfully will be reduced. Much depends on the adoption of more flexible and cost-effective measures, including the incorporation in environmental policy of a system of economic incentives.

As for energy supply, the agenda is straightforward, even if the outcome is not assured. The accommodation of energy and environmental goals can be promoted by encouraging policies and programs that aim at—or are at least consistent with—the following:

• Rendering plentiful fuels, such as coal, less harmful
• Increasing efficiency in the use of a versatile fuel, such as petroleum, so that more work can be done for the same environmental cost
• Making undeveloped but benign energy sources, such as various forms of solar energy, readily available at a reasonable price
• Pursuing research on potential sources, such as fusion, that may eventually combine the promise of abundance with the likelihood of environmental acceptability.

Cleanup Costs and Benefits

How much is spent every year for pollution abatement and how much improvement has been achieved in environmental and human health? To this obvious and highly important question, there is not a very satisfactory answer. Estimates of expenditures vary widely, owing to differences in the way that various analysts select their samples of firms and prepare their estimates. As for estimating benefits, the task is especially difficult because it requires putting dollar values on such difficult-to-measure magnitudes as reductions in illness and the enjoyment of recreational facilities that have been maintained or reclaimed. Several studies suggest that all factors considered, benefits of environmental expenditures have exceeded costs, but they do not explore the question whether these benefits could have been achieved at a much lower cost; nor do they identify which control expenditures have paid off well and which have not.

An examination of available data makes one point abundantly clear, however: As a percentage of the GNP, the annual additional expenditures required to meet government requirements are relatively small—about 1.5 percent a year—but for industries that are most affected, such as petroleum and electric utilities, hundreds of millions are spent annually and account for a substantial proportion of investment expenditures.

The Council on Environmental Quality has prepared estimates of pollution control expenditures (capital costs and operating costs) that are made in response to federal regulations. In 1979 about 60 percent of the total was spent for air pollution control, primarily to curb pollution from energy use. The figures below, in *billions* of 1979 dollars, illustrate how CEQ expected expenditures to rise over the decade.

Program	Actual 1979	Projected 1988	Total 1979–88
Air pollution control			
Private transport	8.1	14.7	115.8
Electric utilities	8.4	13.3	105.0
Other air pollution	5.8	9.6	78.3
Water pollution control			
Electric utilities	.7	1.2	9.4
Other water pollution	12.0	23.2	160.3
Land reclamation	1.4	1.5	15.3
Other (toxic substances, waste)	.5	5.5	34.4
Total expenditures	36.9	69.0	518.5

Source: The CEQ estimates are from the CEQ's Eleventh Annual Report (Washington, D.C., GPO, 1980) p. 394. For a discussion of benefit estimation, see A. Myrick Freeman III, *The Benefits of Environmental Improvements* (Baltimore, Md., Johns Hopkins University Press for Resources for the Future, 1979).

The *Survey of Current Business,* published by the Department of Commerce, estimates pollution abatement expenditures in its national income and product accounts. For 1981 it reported that expenditures for pollution abatement totaled $60 billion in current dollars.

Public Opinion and the Environment

What does the public think about the seriousness of environmental problems and the nation's progress in dealing with them? A number of conclusions emerge from polls conducted by Robert Cameron Mitchell of Resources for the Future and from Mitchell's analysis of other public opinion polls in the past decade:*

- The environment is an enduring public concern along with such issues as defense and health care, but its ranking among the top dozen issues has declined since the late 1970s as the public has become more worried about inflation and unemployment.
- Efforts to improve the nation's environmental quality may have arrested some deterioration but have not brought about any real progress. Nearly as many people think that the situation has worsened slightly (25 percent) as think it has improved (27 percent), and nearly half see no change.

Responses to some selected questions, paraphrased here from 1980 surveys, are given below.

Are we spending too little or too much on environmental problems?	Too little	48%
	Too much	15%
An endangered species must be protected even at the expense of commercial activity.	Agree	73%
	Disagree	20%
What is your attitude toward the environmental movement?	Sympathetic	62%
	Unsympathetic	4%
Should the government screen new chemicals before approving them or wait until a problem develops before taking action?	Screen first	83%
	Wait	8%
What issues worry you a great deal?		
Hazardous waste disposal		86%
Reducing water pollution		54%
Reducing air pollution		36%
Reducing unnecessary noise		11%

Similar results from polls taken in 1981 were reported in a roundup published in *Public Opinion* magazine (February–March 1982, pp. 32–38). For example, 70 percent of the respondents to one poll said that environmental protection laws have either struck the right balance or have not gone far enough; 21 percent said they have gone too far. Commitment is not always matched by knowledge: 80 percent of respondents said they favored regulation of air quality standards, but to a follow-up question, 31 percent said they knew nothing about the Clean Air Act and 39 percent said they knew only "a little."

*Data are from *Environmental Quality 1980,* Eleventh Annual Report of the Council on Environmental Quality, Appendix A: "Public Opinion on Environmental Issues."

BIBLIOGRAPHICAL NOTE

For a dozen years—from 1970 to 1981—the principal source of current environmental information was the Annual Report of the Council on Environmental Quality, published by the Government Printing Office. This chapter draws heavily on those reports as well as on studies conducted at Resources for the Future for the past twenty years, particularly, *Pollution, Prices, and Public Policy,* by Allen V. Kneese and Charles L. Schultze, prepared in cooperation with the Brookings Institution (Washington, D.C., The Brookings Institution, 1975); and Part IV of *Energy in America's Future,* by Sam Schurr and others (Baltimore, Md., Johns Hopkins University Press for Resources for the Future, 1979). The Kneese–Schultze study is particularly useful in explaining concepts and analyzing questions of costs.

A comprehensive treatment of environmental issues appears in Paul R. Portney, ed. *Current Issues in U.S. Environmental Policy* (Baltimore, Md., Johns Hopkins University Press for Resources for the Future, 1978); and William J. Baumol and Wallace E. Oates, *Economics, Environmental Policy, and the Quality of Life* (Englewood Cliffs, N.J., Prentice–Hall, 1979). An excellent chapter on "Managing Air Pollution" appears in Hans Landsberg and coauthors, *Energy: The Next Twenty Years,* a study sponsored by the Ford Foundation and administered by Resources for the Future (Cambridge, Mass., Ballinger, 1979).

To fill the gap left by the discontinuation of the detailed annual CEQ reports, the Conservation Foundation published *State of the Environment 1982* (Washington, D.C., The Conservation Foundation, 1982).

A compendium of charts and tables, covering not only specific environmental changes but also demographic and economic developments affecting the environment, was prepared by the Council on Environmental Quality under the title *Environmental Trends* (Washington, D.C., GPO, 1981). For a readable and thorough treatment of the risk question, see William W. Lowrance, *Of Acceptable Risk: Science and the Determination of Safety* (Los Altos, Calif., William Kaufmann, Inc., 1976).

References to the National Academy of Sciences studies are from *Energy and the Fate of Ecosystems,* Supporting Paper 8, Study of Nuclear and Alternative Energy Systems (Washington, D.C., National Academy Press, 1980); and *Atmosphere-Biosphere Interactions: Toward a Better Understanding of the Ecological Consequences of Fossil Fuel Combustion,* a report prepared by the Committee on the Atmosphere and the Biosphere, National Research Council (Washington, D.C., National Academy Press, 1981).

CHAPTER 7

Energy
in an Unstable
World

The last decade had the global economy reeling from energy shocks. The oil-price explosions of 1973–74 and 1979–80 exacted hundreds of billions of dollars in lost economic output, led to large-scale unemployment, and contributed to a massive increase in international indebtedness on the part of industrializing Third World countries. Even with that experience painfully brought home, energy shockproofing efforts by individual countries or groups of countries can go only so far. The simple fact is that neither the United States, nor any other country, can ever become as immune to external energy developments as it may wish to be. Assume—unrealistically, but for argument's sake— that the United States could produce all the energy it needs. Ignoring the substantial economic costs, would self-sufficiency insulate the nation from events outside its borders? A few examples can point up the futility of such an expectation:

- A major disruption in Persian Gulf oil production could oblige the United States to share domestic energy supplies with its import-dependent allies. The economic consequences of such a disruption would spare no one.
- As buyers, suppliers, and providers of funds, American business firms and banks have a significant stake in energy-related transactions occurring outside the United States itself.

- Committed, as a matter of economic self-interest and national policy, to fostering economic progress in developing countries, the United States has an interest in preventing energy constraints from arresting such progress.

- As the world's major coal exporter, the United States is significantly influenced by energy policies and trends abroad that affect domestic coal prices, coal-mine investments, and the fortunes of ancillary American industries.

- Both as an exporter of commercial nuclear equipment and materials and as a military superpower anxious to suppress the spread of nuclear weapons, the United States has a profound concern with ensuring that civilian nuclear programs abroad do not lead to such weapons proliferation.

The list could be extended. Energy ties between the Soviet Union and Western Europe raise strategic issues affecting the United States. One country's energy production and use can contribute to long-term global environmental problems. But there is no need to belabor the point. Even without its own need for imported energy, the United States is unavoidably drawn into the web of international energy developments and policies. If, in addition, the country remains dependent on oil imports for at least some significant fraction of its requirements, that global link becomes still more important. Some of these worldwide energy interconnections are explored on the pages that follow.

GEOGRAPHIC PATTERNS OF SUPPLY AND DEMAND

At the start of the Industrial Revolution, fuelwood and water power were the predominant energy sources. These two resources were exploited where they were found—the first, because of the impracticability of long-distance shipment; the second, because of the impossibility of it. As a consequence, countries had no choice but to be self-reliant in energy. As coal—followed by oil and natural gas—came into the picture during the nineteenth and twentieth centuries, perpetuation of such self-reliance would have been bought at an intolerably high cost. Mineral fuels, no matter where produced, had become a necessity for industrial development. Unfortunately, the world's bounty of mineral fuels—not only oil and natural gas but coal as well—is conspicuous in its uneven geographic distribution. (See Table 7-1 and, more specifically, the discussion in Chapter 3.) Countries lacking exploitable resources must therefore import fuels from energy-surplus countries in exchange for exportable goods and services.

Table 7-2 shows comparative patterns of energy export propensity or import dependence for selected leading producing or consuming countries. It

TABLE 7-1. Concentration of World Energy Resources, 1980

| | Percentage share accounted for by: | | | | | |
| | Five leading countries | | | Ten leading countries | | |
Energy resource	In reserves	In production	In exports	In reserves	In production	In exports
Coal	76	68	78	93	87	96
Crude oil	57	62	56	75	77	81
Natural gas	72	79	81	83	88	96

Sources: Oil and Gas Journal, December 28,1981; and U.S. Department of Energy, Energy Information Administration, *International Energy Annual, 1980* (Washington, D.C., 1981).

is immediately apparent that the United States, though no longer as self-reliant in energy as it was thirty years ago, is far less dependent on energy imports than such major industrial countries as Japan, West Germany, France, and Italy. Overall, in 1979, the United States was able to meet about 80 percent of its energy requirements from domestic sources, whereas Japan was able to meet only about 8 percent of its needs domestically. More recent trends show the United States with a still less conspicuous degree of import dependence: in 1981, 87 percent of total U.S. energy use was provided by domestic sources, while 13 percent came from net imports. It is important to keep in mind, however, that underlying such an aggregate figure for the United States is substantial dependence on imports for petroleum (more than one-third of U.S. consumption in 1981), offset by a healthy export market for coal (13 to 14 percent of production in the last several years).

In spite of the relatively modest dependence of the United States on energy imports, the country's role in international energy markets—particularly in the oil trade—is a matter of intense preoccupation, both domestically and around the world. There are two reasons for this. First, U.S. imports have risen sharply in recent decades: up until the early post–World War II years, the United States exported more petroleum than it imported; by 1981, as indicated above, oil imports (at a volume of 6 million barrels a day) had risen to roughly 36 percent of consumption. True, 1981 imports represented a scaling-back from an earlier peak—the nearly 9 million barrels per day imported in 1977 accounted for almost one-half of domestic oil consumption. And the picture improved further in 1982, when imports declined still further—to 5 million barrels per day or 33 percent of consumption. But even if we assume modest U.S. oil consumption in the future (as discussed in Chapter 2), our review of the U.S. resource picture (in Chapter 3) points to considerable uncertainty on the energy *supply* side. In short, the United States would be ill-advised to ignore the possibility of continued—perhaps even rising—oil shipments from unstable regions of the world.

TABLE 7-2.   Relative Importance of Energy Exports and Imports
for Selected Countries, 1979

Net energy exporters	Net energy exports as % of energy production
United Arab Emirates	97
Saudi Arabia	94
Libya	93
Kuwait	90
Indonesia	79
Venezuela	76
USSR: total energy	15
oil	29
Canada	8
China	2

Net energy importers	Net imports as % of consumption	
	Total energy	Oil[b]
Italy	98	100
France	92	99
Japan	92	100
West Germany	64	94
United States:		
1979	21	46
1981	13	36
United Kingdom	13	19
Canada	[a]	13

Note: Consumption includes bunkers. All figures apply to 1979 except for the 1979 and 1981 U.S. data, as shown.

Sources: U.S. figures from Monthly Energy Review, May 1982; for other countries, total energy figures from United Nations, 1979 Yearbook of World Energy Statistics (New York, United Nations, 1981); and U.S. Department of Energy, Energy Information Administration, 1980 International Energy Annual (1981).

[a] Canada is a net exporter of energy in the aggregate, as shown in top panel of table.

[b] Because they are unadjusted for inventory change, figures in this column may be slightly off. (Trifling amounts of oil are actually produced in Italy and Japan, as well as in France.)

Second, U.S. imports dominate the international oil trade *in absolute terms* even if not by the relative measures of dependency shown in Table 7-2. America's imports of 6 million barrels a day in 1981 represented 21 percent of worldwide oil flows, and exceeded by 1.5 million barrels a day the total inflow of the second leading importer in the world, Japan. The weight thus exerted by the United States on world oil supplies and, in some circumstances, world oil prices, is a recurrent source of anxiety to countries more acutely dependent on imports than is the United States. Their anxiety is increased if, unimpressed by trends in the early 1980s, they believe that U.S. needs will mean increased pressure on limited world oil supplies in the future.

THE INTERNATIONAL POLITICS OF OIL

World oil transactions, probably much more than trade in other commodities, occur in a political as well as economic context. And that political aspect is not a recent phenomenon. Even before the establishment of OPEC in 1960, the terms which multinational companies negotiated with the oil-producing countries—particularly of the Middle East—were based on more than economic criteria. They were drawn up with the clear-cut involvement of the U.S. government, in pursuit of its own long-term political interests in the producing regions. For example, in 1950, the U.S. State Department—recognizing Saudi Arabia's strategic value to the United States—masterminded an ingenious arrangement, whereby ARAMCO royalty payments to the Saudis could be used to offset the company's U.S. tax liability. (ARAMCO is the group of American companies—Exxon, Standard Oil of California, Texaco, and Mobil—lifting the bulk of Saudi Arabia's oil. Formerly a concessionaire, ARAMCO is now a purchaser.) Politicization of oil goes back a long way.

The formation of OPEC, and events since its inception, brought three jarring elements to the fore in world oil markets: efforts to control supply and thereby price; the role of the Arab–Israeli conflict; and Middle East instability in general. Let us briefly look at each of these three issues.

The new bargaining power of producing countries manifested itself in the attempt to dictate prices and manage supply and in the withdrawal of most oil-exploration and production concessions, which up to then had been awarded to the international companies. No longer did the companies exercise autonomy in critical investment and production decisions. To be sure, the private firms remained important as oil field developers—working under contract—and as major oil purchasers. But the ending of the concessionary era strengthened the hand of the producer countries and was a necessary, if insufficient, step for OPEC success.

OPEC's lasting effectiveness as a cartel is in dispute. (A more detailed account of OPEC's structure and evolution appears in Chapter 5.) Those hesitant to write its obituary can point to sharply rising world oil price increases during the last decade (Figure 7–1) and argue that those increases by and large survived the price cutting and overproduction evident during the 1982 oil glut. Those more skeptical of OPEC's ability to endure cite sharp producer rifts in supply strategy and pricing decisions—illustrated, for example, by the very same oversupply and price-softening conditions of 1981–82 (see box, page 185). They also point to energy conservation, fuel shifting, and rapid growth of oil production outside of OPEC as developments likely to undermine OPEC's

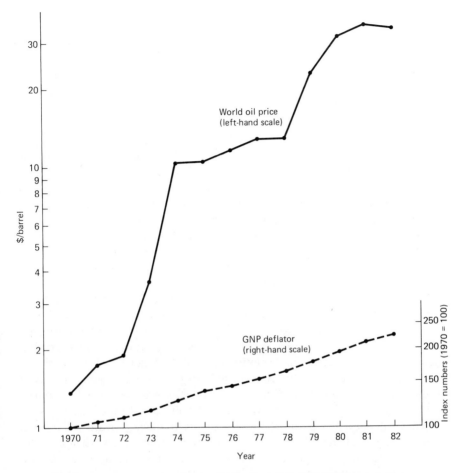

Figure 7-1   World oil price and general U.S. price index, 1970–82. The gap between rapidly rising world oil prices and the overall U.S. price level did not cease its ever-widening trend until the early 1980s. Oil prices are from *Petroleum Intelligence Weekly* (April 12, 1982). Prices refer to official quotations for 34-degree Mideast crude. The GNP deflator is from the *CEA Report* (February 1982); and *Survey of Current Business* (February 1982). Both 1982 figures are provisional estimates. The figure employs ratio scales.

ability to develop and sustain the cohesion of an effective cartel (see Table 7-3). Clearly, much depends on the importance one attaches to the different developments.

Only in retrospect will we be able to tell who is right. Different premises—all plausible—point to different outcomes. Saudi Arabia—whose gigantic oil fields have given it the preeminent voice within OPEC—may be content to deplete its resources only slowly. The Saudis might be confident that alternative energy sources or ''backstop'' technologies—those, such as nuclear

TABLE 7-3.   The Importance of OPEC and Oil, for 1972 and 1981

	1972	1981
OPEC oil as percentage of world oil supplies	52	39
Oil as percentage of world energy consumption	46	42

Source: British Petroleum Company, BP Statistical Review of World Energy 1981 (1982).

or oil shale, which might come into play on a significant scale if oil prices should rise too much—will be chronically handicapped in their growth. Alternately, a Saudi Arabia fearful of permanently losing markets to competing energy sources and hard pressed, as a result, to sustain the tide of domestic expectations engendered by the dramatic growth of its wealth during the past fifteen years, will behave differently. It will schedule production and prices so as to maximize its revenues in a fairly certain short run rather than a problematic long term.

The second potentially destabilizing element on the world oil scene—heightened during the last decade—is the unresolved Arab–Israeli conflict. The Arab members of OPEC, who in 1979 accounted for 70 percent of OPEC oil exports and nearly half of total world oil exports, were joined by several other Arab countries in an embargo of oil shipments to the United States during the 1973 Yom Kippur war and have threatened, intermittently, to embargo shipments again if an "acceptable" solution to the Palestinian issue is not achieved. They look to America as the country able to press for that goal. The Arab threat may or may not be credible. For example, it was not carried out during Israel's 1982 incursion into Lebanon. But the airy dismissal of ten years ago ("What will they do? Drink the oil?") is less frequently heard today, even though the use of the oil weapon, much less its effectiveness when used, remains uncertain. (As we have noted, holding back production for a more opportune time continues to be an alternative to "drinking it" for at least the wealthier producers.) Deliberate disruptions aside, some of the region's governments may not be masters of their own fate: turmoil, insurrection, unpredictability, and fragility of the national power base are hallmarks of Middle Eastern states. Plan, design, and political control can crumble with disconcerting suddenness.

This calls attention to the third point: a major portion of the world's energy supplies originates in an area of the world with a record for marked instability—whether within the Arab fold alone or in the broader Moslem group of nations. The overthrow of the Shah of Iran and the Iranian–Iraqi war are merely recent episodes. No energy-importing country in the world can afford to neglect the strong possibility that a major political or military eruption will unexpectedly send another round of economic shock waves through international oil markets.

A Collapse in the World Oil Price—Conceivable? Desirable?

The period in which this book was being written—1981–82—was one of weak oil demand, abundant supplies, and declining real prices, which especially benefited the United States. (World crude oil prices are set in dollars, and because of the appreciation of the dollar in terms of other currencies, the benefit to other countries was a more limited one.) In the wake of falling revenues, which hit some oil-exporting countries harder than others, OPEC's future success at averting a major price decline by regulating output (an outcome that is partly dependent on the actions of *non-OPEC* countries, such as Mexico) was being questioned. Should an oil price collapse occur—and we believe the possibility cannot be dismissed—its consequences depend on who you are and how your expectations regarding future oil prices governed your actions in recent years.

People in oil-consuming countries can only be better off as a result of a price decline. It will allow one to drive more, to divert savings from one's heating bill to other purposes, and in dozens of ways, it means an increase in one's real income. But what about that super-saving gas furnace or other piece of energy-efficient equipment that you installed at considerable expense in expectation of continued high energy prices? That decision, unfortunately, is water over the dam: depending on the extent of your investment, you conceivably might have been better off to have avoided that extra outlay, but once having made the investment, you cannot be hurt by falling energy prices. Your utility bills will just remain lower.

In a wider perspective, however, there are not only gainers but some identifiable losers. Companies, investors, and creditors, the ultimate success of whose projects rested on a rising oil price, could be left holding the bag—that is, could fail to realize the expected rate of return on their investment, or worse. Oil-exporting countries forced to retrench in their development plans may face political and economic turmoil. If such impacts become magnified and cumulative, they could conceivably more than offset the benefits of lower prices accruing to consumers. More subtly, if a drastic fall in the oil price gives rise to expectations of indefinitely depressed prices, the resulting abandonment of conservation impulses and energy security efforts could lead to profound agony, when, as is likely, these expectations are later disappointed. For these reasons, the issue of whether both consuming and producing countries have a shared interest in a deliberate policy of oil price stabilization or whether things are better left to the marketplace is genuinely worthy of debate.

Keep in mind that, even though oil market developments during 1981–83 seem, in retrospect, to have undermined the more ominous future scenarios depicted in the seventies, the most prudent *long-term* assumption continues to be one of increasing cost of energy resources. At least that is our view.

COPING WITH ENERGY INSECURITY

The precariousness of the world petroleum market in recent years and the threat of future disruption oblige the United States and other vulnerable countries to plan countermeasures. For the long run, there is a widely recognized need to diversify away from concentrated dependence on OPEC oil. (Even that strategy invites some thinking through, as the box on page 187 notes.) To deal with precipitous contingencies, ready-response mechanisms available to oil-importing countries include strategic stockpiling, arrangements to share supplies with other countries, and measures to ease severe economic impacts.

Since patterns of energy use cannot easily be adapted to short-term changes in supply conditions, actual or feared disruptions in the international flow of oil—such as occurred in 1973-74 or 1979-80—can translate quickly into enormous price increases. A major way of blunting such impacts—and forestalling them by creating conditions whereby no country (or countries) would be inclined to deliberately cause a disruption—is to maintain an adequate amount of oil in storage. Such stocks can be drawn down to cushion the effects of shortages. Oil companies routinely maintain inventories in order to meet unanticipated surges in demand, but they do not normally carry stocks designed to deal with the consequences of extraordinary developments, such as a military conflict in the Middle East. Companies can no doubt be compelled, through government regulations, to augment their commercial stocks and target some inventories for global emergencies. But in view of the amount of oil that will have to be acquired and held, the storage facilities needed, and logistical problems to be solved, a major role for government is inescapable.

Although their approaches differ, governments in all major oil-consuming countries have large stockpiling programs. In the United States, the Strategic Petroleum Reserve (SPR) is moving haltingly toward a 1989 goal of 750 million barrels, stored in salt caverns on the Gulf Coast of Louisiana and Texas. In the spring of 1983, the inventory, totaling somewhat over 300 million barrels, was being expanded at a rate substantially below the congressionally directed figure of 220,000 barrels a day. We say "moving haltingly" because one of the unfortunate characteristics of U.S. energy policy, at least with respect to strategic stockpiles, has been to allow short-term considerations to undermine strategic objectives. Thus, in 1982, more resolute building up of the strategic reserve—admittedly a costly proposition—was curtailed in the interest of reducing the federal budget deficit.

The cost of oil insecurity need not be borne by government alone. Since stability in world oil market conditions may at best be an interlude between major eruptions, the market price for oil in the United States might legitimately include a "risk premium" reflecting insecurity. The concept of a premium is not terribly elusive: oil is imported into the United States because it serves a useful purpose, and if, at some cost, we can make those imports safer and less risky, it is a cost we should be willing to absorb. An oil-import fee,

Energy, Security, and Detente. . .

A recurrent theme in this book is that domestic or international circumstances scarcely ever allow one the luxury of pursuing energy goals without considering secondary effects or competing objectives. An issue rich in irony has Western Europe seeking to widen its energy choices (a good thing) but only—at least, so President Reagan argued in 1982—by conferring strategic advantages to the Soviet Union (a bad thing). The USSR is handsomely endowed with oil and gas resources, but lacks the capital and industrial capacity to develop these resources on a timely basis without straining other parts of its strained economy. (Indeed, some Western analysts feel the Soviets—modest exporters of oil in the past—may soon need to add petroleum to grain in their overseas shopping list.) Western Europe, for its part, has been trying to shift its energy profile away from undue dependence on OPEC oil.

Therein the basis for the deal that has commanded so much attention in recent years: Western Europe has contracted to import substantial quantities of Soviet natural gas. Partially in exchange, equipment for construction of the 3,000-mile pipeline through which the fuel is to be transmitted is to be financed by Western European bank credits. The Europeans contend that, although the Soviet imports *will* constitute a rapidly rising proportion of their natural gas availability, these supplies are a long way from exposing the importing countries to Soviet mischief, should interruption in the contracted flow be attempted.

West Germany, for example, plans to get one-fifth or more of its gas from the USSR, but this volume would amount to no more than about 4 percent of the country's overall energy consumption—an exposure, according to the Germans, that can be safeguarded by stockpiling and other means. (One might note, incidentally, that in commercial dealings, the Russians' commitment to sanctity of contracts has tended to be faithfully capitalistic.) Not so, insisted President Reagan, as he sought to deny the use of U.S. and U.S.-licensed equipment in the pipeline project, linking that denial to the institution of martial law in Poland during 1981–82. (American wheat, benefiting from tortured administration logic, escaped export restrictions.) Not only, warned the U.S. government, would there be serious risks of energy cutoffs; the capital infusion would remove Soviet pressure to divert resources from the military sector.

And so an energy question broadens into the wider issue of East–West economic relations and détente. Does an economic squeeze diminish Soviet military adventurism or can it provoke even greater belligerence? Each view has its adherents. Then there is perhaps the equally plausible view that sees a continuation of major East–West commercial contacts as unlikely to heighten, and quite possibly apt to rein in, world tension.

calculated to approximate such a risk premium, would (1) strengthen conservation impulses; (2) encourage additional domestic production of energy; (3) reduce—as a consequence of (1) and (2)—imports below levels otherwise prevailing, thereby also easing upward pressure on the world oil price; and (4) provide at the same time the means for helping to finance such safeguards as the petroleum reserve.

There is no virtue, of course, in imposing arbitrarily high import duties just to minimize the inflow of foreign oil. That would expose the United States to intolerable costs and strain relations with our oil-producing trading partners. An oil-import premium, however abstract that notion may sound, is designed merely to bring into the open the real cost—not adequately reflected in the market price—of oil imports. It does not compromise the objective of a flourishing world trading system.

Not all aspects of energy supply disruptions call for a government response; on the contrary, in one important respect—pricing and allocation decisions—nonintervention is likely to be far more fruitful. (Chapter 5 takes a broad look at government regulation of energy markets.) In 1974 and again in 1979, government price controls on petroleum and efforts to allocate limited supplies by region of the country and sector of use were not very successful. Artificially low prices encouraged consumption (and imports) at the very time when a price increase would have signaled the need to conserve. Interminable service station lines in some areas of the United States, and gasoline surpluses in others, were one conspicuous result of this policy. One can imagine a disruption so severe as to make a government role for managing the crisis inescapable. Nonetheless, the lesson of the 1970s—that, however painful, marketplace adjustments are frequently the least onerous to adapt to—should not be forgotten.

Other countries have also had to cope with the problem caused by insecure world oil supplies. The most notable *collective* effort spurred by the 1973-74 crisis was the establishment of a Paris-based consultative and policy-coordinating group, the International Energy Agency (IEA). Closely linked to the Organisation for Economic Co-operation and Development (OECD), the IEA is composed of the major non-Communist industrial countries, most of them significant oil importers. (France, which had visions—later unrealized—of privileged treatment by Arab members of OPEC, stayed out although it proceeded to participate informally in IEA deliberations.)

The IEA produces statistics and analytical studies on energy trends and policies. It runs computerized exercises simulating oil-disruption scenarios. In a sort of annual "report card," it evaluates performance by member countries in conservation and energy resource development. The IEA has, for example, scolded the United States for levying gasoline taxes at a level only about one-third as high as those prevailing in many other countries.

But probably the IEA's most formidable task—which its member countries are pledged to implement—is that of ensuring an equitable sharing of any shortfall in oil supplies induced by a disruption. The IEA oil-sharing plan can be activated in one of two ways: when an individual member country experiences a reduction in its normal oil supply of more than 7 percent; or by a shortfall of that magnitude for the IEA member countries as a group. Affected countries are obliged to meet their deficiency by curtailing demand and drawing on their inventories before qualifying for assistance. When invoked, the triggering mechanism could oblige the United States to share some of its imports with the worst-hit country or countries and, under extreme circumstances, compel the United States to export some of its domestic oil supplies.

Although the IEA evaluates member countries for their commitment to sound energy policies, the IEA itself—and the arrangements it oversees—defy easy evaluation. The trouble is that, as with other international agreements, one is not sure whether domestic interests will override transnational obligations and interests in a crisis situation. The IEA may be nervous about testing the sharing procedure—even when the objective conditions for the triggering action are present—for fear of exposing a paper tiger. For example, during the 1979 oil shortage accompanying the Iranian revolution, Sweden's oil shortfall actually qualified for the IEA sharing scheme. But with some straining, ways were found to hold off on the action plan and thus avoid the test. In this country, since the U.S. obligation stems from an executive agreement rather than a Senate-ratified treaty, it has never been clear whether presidential emergency powers will suffice to implement the program and what role Congress might play. The suspicion is that it would play whatever role it desires.

The IEA partnership can undoubtedly confer selected strategic benefits. But if the record of other multicountry cooperative ventures (with probably more at stake in case of failure) is a guide, different countries will continue to pursue their energy interests largely through national efforts. This is so despite the fact that any sudden competitive scramble in pursuit of purely national interests may leave everyone worse off than would collective action. (For example, if everyone amasses large oil inventories during a crisis, this would cause a rapid bidding up of the price.) But as long as the cooperative approach is expected to be violated by others, success at joint action may prove elusive.

CIVILIAN NUCLEAR POWER AND PROLIFERATION

Civilian nuclear power, perceived some three decades ago as the path to cheap energy, has emerged as a prominent alternative for reduced dependence on oil, but there are fairly severe problems. Other parts of this book (particularly Chapters 3, 4, and 6) look at nuclear power from the standpoint of resource potential, R & D, and environment. Here we are primarily concerned with the weapons-proliferation aspects of nuclear energy. Few issues deserve more thoughtful attention and few pose such soul-searching questions for the United States.

Nuclear power has witnessed many a slip between resolve and achievement. Forecasts in 1971 estimated that U.S. nuclear generating capacity in 1980 would reach 150,000 megawatts (MW); the level actually attained was, in fact, only 53,000 MW. And, in the wake of plant cancellations, deferments, uncertainty, and—not least—a sharply declining growth rate in overall electricity demand, the target once foreseen for 1980 is unlikely to be reached before 1995, if then. Other countries, too, have reined in their nuclear power programs, but not nearly to the same extent. Such retrenchment from previously enunciated projections ought to prompt us not to treat specific goals too seriously. Nonetheless, in spite of halting progress, installed nuclear power capacity around the world has continued to grow at relatively fast rates. In the early 1980s, nuclear power accounted for 10 percent or more of overall electricity-generating capacity in a number of advanced industrial countries, as Table 7-4 shows. (Table 7-4 shows only plants under construction; it omits facilities announced or projected, for such projections, as noted, have turned out to be notoriously fickle.)

Anxiety over the role of nuclear energy is twofold: risks arising from routine operation of atomic power plants and the associated management of the nuclear "fuel cycle"; and the risks—with far graver consequences if borne out—of diversion of nuclear materials for weapons purposes. The latter is what injects the "proliferation" issue into civilian nuclear power development. (The ultimate risk, of course, concerns the possibility of such weapons being used.)

An important question, therefore, concerns the destabilizing consequences of countries intent upon becoming self-reliant in the major stages of the nuclear fuel cycle. One of the tempting features of nuclear power is that it is perceived to confer a substantially greater degree of energy independence in an era when oil imports are at best costly and at worst insecure. That line of thinking, even without entertaining nuclear weapons objectives, sees attractiveness in developing a commercial, indigenous capability for enrichment at the "front end" of the nuclear fuel cycle and reprocessing at the "back end." (More particulars about the nuclear fuel cycle are found in Chapter 4.)

From the standpoint of weapons production and proliferation, the concern with both enrichment and reprocessing is that both stages can be innocent components of a civilian nuclear power program, or constitute a critical juncture for bomb-making. Among the countries listed in Table 7-5, which summarizes the status of known enrichment and reprocessing activities, those with a publicly acknowledged weapons-making capability include the United States, the USSR, the United Kingdom, France, and China. India, with its nuclear detonation in 1974, hovers close to the list. Others, including some not shown, invite greater or lesser conjecture as to where things stand.

Supposedly, one ought to feel more reassured about nonweapon coun-

TABLE 7–4. The Role of Nuclear Power in Selected Countries

Country	In operation, end of 1981			Under construction	
	No. of units	Installed capacity (megawatts)	As % of total electricity capacity[a]	No. of units	Capacity (megawatts)
United States	74	55,051	12	79	87,217
Canada	11	5,494	7	14	9,751
France	30	21,595	34	26	28,585
West Germany	14	8,606	10	10	10,636
Sweden	9	6,415	23	3	3,025
Switzerland	4	1,940	14	1	942
United Kingdom	32	7,627	10	9	5,533
Taiwan	3	2,159	—	3	2,765
Korea	1	564	5	8	6,869
Japan	24	14,994	10	12	9,973
India	4	809	2	4	880
Argentina	1	335	3	2	1,292
USSR	35	14,036	5	25	24,260
East Germany	5	1,694	9	4	1,644
Others	25	12,453	—	38	28,646
World	272	153,772	—	238	222,018

Sources: U.S. figures for first three columns are from Monthly Energy Review, May 1982; denominator for calculating the third column is from UN, Economic Commission for Europe, The Electric Power Situation in the ECE Region and its Prospects, Geneva, December 23, 1981, and UN, Yearbook of World Energy Statistics (New York, United Nations, 1982); all other figures are from International Atomic Energy Agency, Bulletin, March 1982, p. 2.

[a] Except for the United States, the total electric capacity figures on which the percentages in this column are based refer to 1980. Hence, the percentages shown are probably slightly overstated in most cases.

tries that are signatories to the International Non-Proliferation Treaty (1968) barring diversion of nuclear materials from peaceful to military purposes and which subject themselves to surveillance by the International Atomic Energy Agency. (Evidently Israel, who is not a signatory, bombed an Iraqi research reactor in 1981 because it mistrusted Iraq's signatory status as proof of harmless intent.) More uncertainty applies, presumably, to nonsignatories, a list which in mid-1982 included, besides Israel, South Africa, Argentina, Saudi Arabia, Algeria, India, Pakistan, and Brazil.

Conjecture aside, the weapons potential of civilian nuclear power programs is a deeply disturbing one. For the time being—but probably only for the time being—the two stages of the nuclear fuel cycle with serious military implications are largely in the hands of "nuclear club" members or advanced industrial countries. Even in the United States, the prospect of spent-fuel reprocessing and possible development of a plutonium-fueled breeder reactor have aroused deep political controversy. Opponents of such developments hope both to arrest the nation's nuclear power growth and, arguing the greater pro-

TABLE 7-5. **Status of Nuclear Enrichment and Reprocessing Facilities, 1980**

| Country | In operation or operable | | Planned | |
	Enrichment	Reprocessing	Enrichment	Reprocessing
United States	X		X	
France	X	X	X	X
Urenco[a]	X		X	
Japan	[b]	X	X	X
Brazil			X	
South Africa			X	
USSR	X		[c]	
Belgium		X		
United Kingdom				X
Argentina				X
West Germany		X		X
India		X		X
Italy		X		X

Source: Joseph A. Yager, *International Cooperation in Nuclear Energy,* (Washington, D.C., Brookings Institution, 1981) pp. 19–21.

[a] British–Dutch–West German consortium.

[b] Japan possessed negligible enrichment capacity in 1980.

[c] Information unavailable.

liferation potentials of the breeder and of reprocessing, to have such restraint serve somehow as a model for other countries. A question that has not been answered is whether unilateral action in this field by one country can significantly shape the policies of others. In fact, efforts at "internationalizing" the nuclear fuel cycle—which go back to a time, after World War II, when U.S. influence in these matters was much greater—have proved frustratingly elusive.

Another, more practical, question concerns the diversity of routes to acquisition of a nuclear weapons capability. We have dwelt on commercial-scale enrichment and reprocessing, for those are the components of civilian nuclear electric-power programs often singled out as posing the principal proliferation threat. But there are other ways—and less sensational ones than theft or highjacking of weapons-grade material—to achieve that objective. Whether or not countries have a nuclear electric generating capability, a number of them have small research reactors for scientific or medical purposes. These units often require highly enriched uranium which, if diverted, can be deployed for military purposes. Other research reactors, utilizing nonenriched uranium, can nonetheless be the pathway for a crude, small-scale reprocessing effort, yielding sufficient weapons-grade material (plutonium) for fashioning an explosive device. India, with its 1974 atomic explosion, is presumed to have followed this route. Clearly, restraining nuclear-based electricity as an energy option need not end the proliferation threat.

In all this, we must bear in mind, however, that nuclear power and pro-liferation questions do not reduce easily to "good guy–bad guy" camps in which motives are either sinister or innocent. Some of the industrially develop-ing countries, traditionally sensitized by what they feel to be cavalier treatment by the developed countries, chafe under the "second-class" status that statutes like the international safeguards treaty (noted above) and the U.S. Nuclear Non-Proliferation Act (1978) seem to confer on them. That is, their rejection of such measures—which are designed to ensure that only peaceful ap-plications of nuclear power take place in nonweapons states—need not in-evitably betray clandestine military activities, but can reflect genuine resentment at the condescension perceived to be exercised by the nuclear "haves" of the world. (Moreover, it should be noted that some countries' unwillingness to ratify the International Nuclear Non-Proliferation Treaty has not precluded their submission to safeguard inspections by the International Atomic Energy Agency.)

There is no easy way out of such a dilemma. For the United States and other suppliers of nuclear technology, relaxation of insistence upon safeguards in countries receiving such technology would surely be unthinkable. At the same time, the supplier nations can perhaps relieve the anxiety of developing countries eager to develop civilian nuclear power by making it unambiguously clear that, subject to comprehensive safeguards, they will supply nuclear ma-terial, equipment, and technology on a noncapricious basis. (But to avoid capriciousness does not mean that one can always afford to be categorically nondiscriminatory: circumstantial evidence on nuclear weapons planning and activity has to weigh in supplier-nation policies.) Supplier nations can also promote a more stable nuclear environment by offering tangible economic incentives that would encourage developing countries to forego enrichment and reprocessing activities. Finally, though disappointing so far, the search for more proliferation-resistant fuel cycles should not be abandoned.

This part of the international energy story necessarily ends on a fragile note. Traditional appeals to human reason have stressed plowshares at the expense of swords and butter at the expense of guns. The unsettling aspect of nuclear energy is the possibility that one need not be at the expense of the other: electricity and weapons can jointly be produced through nuclear tech-nology. Therein lies the major challenge for those having to fashion national and international institutions and policies for the peaceful uses of the atom in the years ahead.

ENERGY AND THE DEVELOPING COUNTRIES

The promise of civilian nuclear power has stirred the imagination not only of the technologically advanced countries but, as noted, of some of the world's developing countries as well. Even with military applications a mere subliminal thought, these less-developed countries (LDC) have viewed nuclear technology as a route to energy independence and national prestige. In some countries, earlier longings faded as the economics and complexity of nuclear power became apparent to national planners. For example, the minimum reasonable and, equally important, commercially available size for a single nuclear power plant would frequently represent a substantial fraction of a developing nation's entire electricity-generating capacity. Under such circumstances, where a single plant failure could black out a large portion of the country, more modestly scaled power units, employing other fuels, would probably be far more appropriate.

The drawbacks of the nuclear energy option give added weight to the broader problems of energy availability in energy-deficient LDCs. These problems can be summed up in the following series of observations:

1. One can argue about precisely how much energy is needed to accommodate rising living standards and economic development, but much more energy will have to be available to the LDCs in the years ahead to support the kinds of income-growth rates to which many developing countries legitimately aspire. By far the largest part of increases in world oil demand over the next several decades is likely to originate in the oil-importing developing countries.

2. Such needs must be set against the background of the 1970s, when sharply rising oil prices created severe balance-of-payments difficulties for numerous developing countries, gave rise to a ballooning external debt and jeopardized economic growth prospects. As Figure 7-2 suggests, the rise in world oil prices has forced many developing countries to devote an increased fraction of their export earnings to financing fuel imports. Advanced countries have confronted a similar dilemma: U.S. energy imports were equivalent to 9 percent of export earnings in 1960 and to 31 percent in 1978. But for countries highly dependent on a variety of imported goods to sustain their development pace, such a turn of events has been especially burdensome.

3. On the whole, the prospects for easing the distress occasioned by large and growing oil imports are by no means hopeless. The upward oil price pressures of the 1970s could abate. Conservation opportunities exist. Oil and gas exploration has not been carried very far in many developing parts of the world containing geologically promising characteristics. Though there is no certainty that significant quantities will in fact be discovered, the possibility clearly exists. Biomass-derived fuels and solar energy applications offer promise (with special potential for the latter in tropical regions). Hydroelectricity and coal development can ease oil dependence. The challenge is to advance from potential and promise toward realization.

4. The diversity of choices just enumerated looks good on paper; and, as noted, it is not without substance. But the barriers to achieving progress

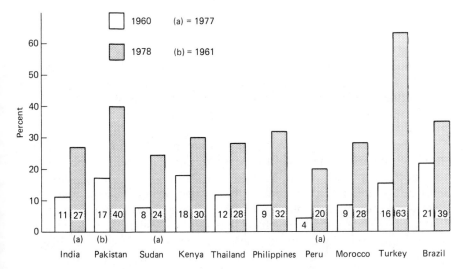

Figure 7-2 Energy imports as percentage of merchandise exports, for selected developing countries, in 1960 and 1978 (unless otherwise indicated). Numerous countries have had to devote sharply rising proportions of their export earnings to finance imports of energy—principally oil. (From World Bank, *World Development Report 1981,* pp. 146–147.)

are formidable. They have to do with government policies and institutions, technical skills and understanding, "infrastructure," and mobilization of capital. Let us illustrate each of these four aspects:

- Rules governing exploration and development can range all the way from confiscatory to accomodating. Subsidized energy prices can deter conservation.

- Innovative resource-development schemes—for example, solar systems—demand technical sophistication, frequently in short supply. Poorly designed or random exploitation of biomass potential can occur at the price of deforestation and reduced agricultural productivity.

- Exploration for and development of conventional fuels may lack the multiplicity of supporting goods and services, such as port facilities and highways, that can mean success or failure.

- Major investments—again, we need not be detained here with fine-tuned estimates—must be made if opportunities of the sort mentioned are to be seized. The World Bank, which has been struggling to become a more effective instrument for multilateral financial support of energy development in LDCs, has figured capital requirements for the energy sector in developing countries at more than $50 billion per year during the next decade, an amount that is twice the recent rate.

The challenge clearly is one that exceeds by far the capacity of the energy-short developing countries, particularly the poorer of the lot. Success depends on a concerted effort involving rich and poor alike. As with development assistance programs in general, there is a clear-cut need for some minimum level of support from the United States and other well-off nations (including the wealthier OPEC members) to strengthen the energy position of the LDCs. Such support does not ensure economic and political stability of the developing world. But it at least contributes to the possibility of that outcome.

The world faces an energy transition of both a short- and long-term character. The immediate need is to find ways to make the global economic system as resilient as possible to recurrent energy shocks. For the long haul, there must be a shared global concern with fashioning energy demand and supply strategies which support the aspirations of people in all developing nations—both oil-producing and others—for economic and social betterment. That goal would be ill served by competitive energy scrambles, military posturing, and efforts to carve out special bilateral deals. It *will* be served by a determined and widespread effort to make the long-run energy transition a smooth and stable one.

Which brings us full circle to the opening thoughts of this chapter: energy is just one more powerful reminder of global interdependence. Recognition of that interdependence can mean the difference between hope and despair in the years ahead.

BIBLIOGRAPHICAL NOTE

A handy annual statistical publication on world energy reserves, production, trade, and consumption is the British Petroleum Company's *BP Statistical Review of World Energy*. Other international energy statistical compilations are issued by the United Nations, the International Energy Agency, the World Energy Conference, and the Energy Information Administration of the United States Department of Energy. The World Bank publishes a yearly *World Development Report* which presents useful text and data on many economic and social trends—among them energy—with special emphasis on developing countries.

Many books have been written on the history of the world petroleum industry. A highly readable introduction is Anthony Sampson, *The Seven Sisters* (New York, Viking, 1975). Periodic reports from the International Energy Agency deal with energy trends and policies on conservation, diversification, and security of supply around the world—for example, IEA, *Energy Policies and Programmes of IEA Countries: 1980 Review* (Paris, OECD, 1981); *Energy Conservation: The Role of Demand Management in the 1980s* (Paris, OECD, 1981); and the comprehensive volume on these and many other global energy questions, *World Energy Outlook* (Paris, OECD, 1982). The energy problems facing less developed countries are discussed by Joy Dunkerley, William Ramsay, Lincoln Gordon, and Elizabeth Cecelski in *Energy Strategies for Developing Nations* (Baltimore, Maryland, Johns Hopkins University Press for Resources for the Future, 1981). A person wishing to delve into worldwide civilian nuclear power issues will be helped by the material in Joseph A. Yager's *International Cooperation in Nuclear Energy* (Washington, D.C., Brookings Institution, 1981).

Suggestions for Further Reading

References to specific topics appear at the ends of chapters. The following is a more general listing of books spanning a broad range of energy topics. While some material in these books demands technical understanding, much of it is accessible to the general reader.

LANDSBERG, HANS H., and others, *Energy: The Next Twenty Years* (Cambridge, Mass., Ballinger, 1979).

NATIONAL ACADEMY OF SCIENCES, *Energy in Transition 1985–2010,* Final Report of the Committee on Nuclear and Alternative Energy Systems (San Francisco, W. H. Freeman, 1980).

SCHURR, SAM H., JOEL DARMSTADTER, HARRY PERRY, WILLIAM RAMSAY, and MILTON RUSSELL, *Energy in America's Future: The Choices Before Us* (Baltimore, Md., Johns Hopkins University Press for Resources for the Future, 1979).

STOBAUGH, ROBERT, and DANIEL YERGIN, eds., *Energy Future* (New York, Ballantine, 1980).

In addition to numerous energy books published in recent years, articles on energy appear in a number of specialized magazines. Many such articles deal with current developments in technological, economic, and public policy aspects of energy. Among such magazines are *Science, Scientific American, National Journal, Technology Review* (published at MIT), *Bulletin of the Atomic Scientists, Foreign Affairs, The New Scientist,* and *Petroleum Econ-*

omist (the last two published in Great Britain). The *Annual Review of Energy* (published in Palo Alto, Calif.) is a collection of somewhat longer pieces on a variety of topics. The *National Geographic* magazine features occasional energy articles; a special (February 1981) issue was devoted exclusively to energy.

Some membership organizations issue publications on energy emphasizing explanation of principles, clarity of presentation, and broad access. Notable examples are the League of Women Voters and National Science Teachers Association (both in Washington, D.C.).

There are a number of specialized research institutions and facilities many of whose publications deal with energy. Examples include Oak Ridge National Laboratory (and other U.S. national laboratories), Massachusetts Institute of Technology's Energy Laboratory (Cambridge, Mass.), Resources for the Future (Washington, D.C.), the RAND Corporation (Santa Monica, Calif.), and the Institute for Energy Analysis (Oak Ridge, Tenn.).

Living with Some Chronic Issues

Energy involves a bundle of chronic issues—the ones that will not go away because there is no "solution." Key congressional votes or regulatory decisions on these questions are made, not once per decade, but every two or three years. Any of them are capable of appearing in newspaper headlines tomorrow. Like it or not, living with the energy problem means living with the uncertainty surrounding these issues and cutting through debate that is unusually muddled with confusing rhetoric. The selective statistic, the slanted interpretation, and the exaggerated accusation are the tools of persuasion. As a result it is difficult to reach the sensible compromises in which all parties give a little so some progress can take place.

This book was written in the hope that better understanding of these issues will help break the logjam obstructing much of energy policymaking. As people think through their positions, perhaps they will reject the "take no prisoners" approach. With this in mind, we prepared an Epilogue devoted to several of the most stubborn issues. For each, a brief overview is provided of different perspectives which play prominent roles in framing the debate. In so doing, there is an effort not to let the authors' leanings (which in some cases are evident from earlier chapters) tilt the discussion toward one perspective and to avoid reducing viewpoints to caricatures. Since most of the issues have been discussed previously, no effort is made to provide background information.

The overviews are followed by short lists of questions. These lists do not include general questions relevant to almost any of the matters raised. For example, a reader's judgments on these questions will undoubtedly reflect an opinion on what represents the proper role for government in economic, political, and social affairs generally. How active should government be in protecting the environment? Should government use tax and spending policies to shape consumption and production trends? To what extent is it the United States' business to worry about the rest of the world? These are broad questions which go to the heart of an individual's political philosophy.

The questions we will raise are more narrow. Some call for judgments which should help readers make other decisions on the larger issues. Others point to areas where readers may want to enhance their knowledge. Still others are designed to call attention to the weaker parts of the arguments presented and to the difficult tradeoffs involved in these matters. By and large, the issues are approached as they typically are addressed in public discussion, reflecting many perspectives apart from those presented thus far in this book.

REGULATING OIL AND GAS PRICES

Four arguments are commonly made in support of regulated oil and gas prices. First, that price controls are necessary to protect low-income households which, on average, spend a disproportionately large percentage of their income on energy; second, that price controls reduce the rate of inflation by holding down the price of a key commodity; third, that price controls prevent energy companies from reaping large, unearned or "windfall" profits; and, fourth, that higher prices will lead to little or no increase in supply. In this view, the large energy corporations have tremendous discretion about pricing. Low-income households, unable to quickly change their consumption patterns, are at the firms' mercy unless government holds prices at a reasonable level. Proponents of price controls argue that, at the very least, price controls can ease the transition to higher price levels by slowing down the sharp price increases which are most damaging.

Critics of price controls counter that markets for oil and gas are reasonably competitive, that established firms must fight for business and that new firms are free to enter the market. Because of competition, there are no enduring windfall profits. They say prices need to be free to reflect changes in supply and demand if society is to use its resources efficiently, and that suppressed prices lead to excessive consumption and insufficient production. The effect of high energy prices on particular groups is not the issue because, in this view, the distribution of income should not be the concern of energy policy. Society is better off if prices are set at a level reflecting consumers' willingness to pay and suppliers' cost of production and ability to compete.

Because price controls encourage consumption by keeping prices artificially low, they increase the need for imports. But a higher level of imports will have negative effects on the country's balance of payments and leave the United States more vulnerable to the whims of foreign suppliers. Energy will be consumed in uses where its value is less than the value of the goods and services shipped abroad to buy it. Price controls also retard the emergence of alternative energy sources.

A third perspective is characterized by opposition to price controls and support of direct aid to individuals hit hardest by higher energy prices. In this way, the negative effects of price controls on the economy are avoided while the impact of high energy costs on the poor is partially offset. Although taxes will pay for these programs, the cost to society will be less than if prices are regulated. In addition, the political pressure to apply price controls will be eased. Those who favor controls respond, however, that direct aid programs have typically been poorly funded. They fear that political problems will always prevent the creation of aid programs of sufficient size to deal with the problem.

Questions To Think About

- Assuming that price controls create less efficient markets, are they nonetheless preferable to making consumers, and particularly low-income consumers, bear the brunt of higher energy prices?
- Does the fact that energy is a basic need for everyone suggest that market pricing is inappropriate? In this respect, is energy a special case?
- How well have oil and gas price controls worked in the past? Can they be designed to work more efficiently? (See Chapter 5 for a discussion of the evolution of price regulations.)
- What evidence is there that oil and gas markets are competitive? Chapter 5 can offer advice on indicators to look at—for example, profit margins, portion of the market dominated by large firms—and how to evaluate them.
- Can consumption of energy be significantly reduced under price controls without complex regulations and possibly severe economic problems?
- Does the political process work against energy aid programs? One indicator is the level of funding for the federal government's Low Income Energy Assistance program, the largest program designed to help poor individuals pay high energy bills.

MINIMIZING RELIANCE ON IMPORTED OIL

Since the oil embargo of 1973, statistics on oil imports have defined the success of American energy policy for many people. But there is disagreement over what should be done. One view is that swift, decisive measures to reduce imports significantly are necessary. It is pointed out that as the United States imports more oil it becomes more susceptible to restrictions on its diplomatic actions, reduces the amount of oil available to other importers, and puts upward pressure on world oil prices. Thus, the best response is to develop alternatives to foreign supplies. This leads to policies such as subsidies to synthetic fuels production or other energy sources that have import-reducing potential. Another route to import reduction is through tough quotas. Such quotas have the added benefit of assuring domestic energy producers of a portion of the market. With this access, new domestic energy industries that will eventually offer an economical alternative to imports will develop and established industries will thrive.

Critics of proposals for drastic reductions in imports argue that energy independence is an illusory goal, since the U.S. economy would still be linked to the rest of the world and other industrialized countries would remain large oil importers. They also note that imports are cheaper or more environmentally benign than what would emerge from the large-scale application of most of the alternatives suggested. Therefore, a move to a position of zero oil imports will require either massive subsidies or the scaling back of environmental goals. Instead, they recommend steps to enable the United States to continue to import oil but at a lower risk. In this way, they say, the link between dependence on oil imports and vulnerability to an oil supply disruption can be weakened. These measures could include stockpiles or moderate tariffs, whose size would be determined by careful calculation of the risks of imported oil.

This view is logically accompanied by belief that market forces, not government directives, should determine import levels. If domestic energy alternatives are more costly than imports, it is argued, then society needlessly wastes resources when it consumes the domestic alternatives. This is true even if prices of the domestic energy source are lower due to a subsidy since the subsidy will be paid by taxpayers. This view is not incompatible with a tariff on oil imports, provided it can be shown that imports carry external costs not reflected in their price. In this view, it is preferable to reduce imports by using taxes to raise their price rather than limiting the quantity of foreign oil purchased.

Questions To Think About

- How do the costs of current programs, such as stockpiles or subsidies to domestic energy producers, compare with the possible costs (at some date in the future) of an oil supply disruption?
- When alternatives to oil imports are proposed, are the costs of those alternatives spelled out? Is the cost of potential damage to the environment included in these calculations?
- How large should the U.S. oil stockpile be in order to reduce significantly the impact of a disruption of oil imports (see Chapter 7 for a discussion of stockpiling)?

- How likely is an oil import disruption? How does the amount of oil likely to be removed from the market by a given political event compare with that which would be available from spare capacity and petroleum stockpiles?
- Are useful emergency contingency plans in place or do political factors make the development of such plans impossible? (More about this below)
- Does the concern about the dangers of oil imports assume that oil exporters can cut off the flow of oil at will, without regard for their need for revenue?

MANAGING EMERGENCY ENERGY SHORTAGES

The debate over the response to energy emergencies is a debate over priorities. Some see allowing markets to adjust effectively to higher prices as the first priority. Others want to make sure income distribution is not skewed by the price change. Still others call for protection of important sectors of the economy.

Those who want to leave prices free to rise oppose plans to have the government ration scarce supplies out of belief that such large-scale government intervention will damage the economy. They claim allocation and rationing programs will inevitably lead to inefficient use of petroleum. Rationed supplies would be available for those with the most political power and not for the areas of the economy where they can be used most productively. Free market proponents argue that if prices are not regulated, those with the greatest need for petroleum products (as reflected by their willingness to pay higher prices) will be able to get them. In this view, the cost of softening the impact of higher prices on some consumers through rationing is unacceptably high since most consumers will be worse off (see Chapter 5 for more on rationing).

Some of the critics of price controls and gasoline rationing do not, however, counsel a pure "hands off" policy. One alternative is to increase government payments to needy individuals during an energy emergency. In addition, some would like government to impose taxes or tariffs on oil to reduce consumption with the revenues being rebated back to consumers. By restraining demand, these taxes or tariffs would make it more difficult for oil exporters to raise prices. Oil from stockpiles could be used to further reduce the gap between supply and demand.

But such programs are unsatisfactory responses to those whose primary concern is that low-income groups are not crushed by high energy prices, especially with the skepticism about the political viability of large energy aid programs for the poor. They push for a limit on price increases, arguing that allowing prices to rise to levels set by the market during a shortage will leave the poor unable to buy what they need. It is also noted that a sharp rise in energy prices will divert the income of consumers toward unavoidable energy purchases, depressing other areas of the economy. They favor gasoline rationing, designed to assure everyone a minimum amount of the rationed product at a controlled price.

Support for allocations also comes from those who fear that a free market will seriously damage certain critical sectors of the economy. For this reason, there is support for programs to provide access to "reasonable levels" of price-controlled petroleum products for certain groups through some sort of allocation system. Typical candidates for special treatment are farmers, energy producers, and police and fire departments.

Questions To Think About

- Will the clamor of special interest groups for government intervention inevitably be successful once an emergency hits? If so, does this argue in favor of having some plan for government intervention ready in advance?

- Will a standby plan of price controls and rationing (or belief that such a plan will be imposed) take away the incentive for businesses to prepare for disruptions through actions such as increasing their oil inventories or preparing to switch to other sources of energy?

- Are there sensible and practical criteria for rejecting one industry's claim for special access to price-controlled energy but accepting another's?

- Do advocates of taxes and tariffs have a workable plan detailing how those revenues would be quickly returned to consumers or otherwise used to alleviate higher costs?

- Does the government have a plan for using oil from government stockpiles in an emergency? What will it mean for prices, consumers, and the economy as a whole?

DEVELOPING ENERGY ON FEDERAL LANDS

What the federal government does with the land it owns but has not yet opened for exploration and development is the subject of debate between those who worry the environment will be seriously damaged and those who value more highly the benefits from increased production of energy resources. The debate also involves the question of whether a resource produced now will do more for the United States than a resource produced at a later time.

One view is that, as a general rule, concern for the preservation of the environment should take precedence over the gains from increased energy production on federal lands. The value of preserving these areas is obvious from the increasing numbers of hunters, campers, and tourists who visit them. Some of these lands, if disturbed, can never be returned to their original state. As a consequence, leasing and sales of federal lands should proceed at an extremely cautious pace with every safeguard utilized to prevent environmental problems such as the destruction of wildlife habitats. In some particularly scenic spots, such as official wilderness areas, no drilling or exploration should be permitted.

Another view is that the value of energy resources on federal lands and offshore is potentially large. If there is no production in these areas, more expensive ways of meeting the energy needs of society will be sought, with higher energy prices and lower economic growth resulting. Nor is it evident that energy exploration and production will necessarily cause unacceptable environmental damage. A host of environmental regulations will apply. And those who favor increased leasing cite several successful industry-sponsored programs to reclaim damaged land.

Different views are also held on the question of whether resources should be saved for the future. To some, exploration and production now will only lead to shortages and higher prices later. The value of these resources will be much greater for later generations than for the present one, something which corporations overlook in their pursuit of short-term profits. Thus it is particularly short-sighted for government to offer leases for lands which contain minerals that are in great abundance, such as coal. Production from such lands will do little to lower general price levels and will net the government only small returns.

But defenders of accelerated development point out that private companies have a financial interest in being aware of the long-term value of their resources. A sensible producer, knowing net returns will be higher at a later date, will wait. However, if sales are more profitable now, there will be no need to delay production. In this way, the market will account for the long-term scarcity of the resource, assuming that prices are allowed to rise to reflect that scarcity. Prospects for the future are enhanced as later generations have a larger stock of goods and a greater productive capacity. Thus, leasing land for mineral exploration makes sense even if supplies are already abundant. Some of the deposits on federal land may have lower costs or superior location. Their production could contribute to lower price levels.

Questions To Think About

- Should the strong interest of those concerned about damage to scenic lands dominate the less intense interest that all energy consumers have in the slightly lower prices that would result from moderately increased production?

- Do energy corporations place too much emphasis on short-term profits at the expense of the long-term picture? Is the result likely to be future shortages of important energy resources?
- How well can environmental goals be protected through regulations or by providing economic incentives without prohibiting energy exploration and development activities (Chapter 6 provides a discussion of these issues)? What is the past record?
- Do private interests have the proper incentives for taking sufficient account of the environmental impacts their actions may have? Are there ways they can be instituted where incentives have been inadequate in the past?
- Should the government wait until energy prices are high and market conditions tight before offering and setting terms for leases on public lands?
- Can one trace the impact on either the environment or in terms of energy production, of accelerated offshore leasing?

MANAGING ENERGY-RELATED ENVIRONMENTAL DANGERS

The conflict between energy and environmental goals revolves around disagreement over a critical question: What is an acceptable level of environmental damage from energy production and how can that level be achieved at the lowest cost?

From one perspective, only a low level of energy-related environmental damage is viewed as acceptable and government regulations should be used to restrict or modify energy-related activities which threaten this level. The benefits of many energy activities are outweighed by their environmental costs unless those activities are tightly regulated. In this view, coal mining and burning, synthetic fuels production, and offshore oil and gas drilling are a few of these areas. The stakes are so high (especially because of potential damage to human health) that there is little room for flexibility. Regulations should be based on the best scientific knowledge about the levels beyond which pollutants will cause serious damage. Polluting companies should be forced to use the most effective pollution-control technology.

Others would place less value on environmental goals and greater value on the benefits of increased energy production and economic growth. They also frequently question the connection between the energy-producing or -consuming activity and the asserted environmental damage. From this viewpoint, there is opposition to some regulations and a desire to assure that others are flexible enough to reflect the uncertainty surrounding the issue.

Those who favor reduced regulation also tend to support use of cost-benefit analysis to shape policy. They feel their position will generally be strengthened if costs and benefits of proposed regulations can be translated into dollars and cents. On the other hand, many environmentalists are suspicious of these techniques, pointing to the difficulty of attaching numbers to, or "monetizing," benefits of clean air and water and especially the value of a human life.

A third perspective is provided by those who propose to rely on market mechanisms to reduce pollution by such means as emission fees that penalize polluters. They believe conflict between economic growth and environmental damage is not properly handled by regulations placing uniform limits on emissions or requiring use of a particular technology. First, such regulations place the same restrictions on emissions which are expensive to control as on those which are cheap. Second, no incentive is offered to develop innovative ways to reduce pollution. Instead, polluting companies should pay fees based on the amount of pollution and what it would cost to clean it up. Companies could assess whether the costs of continuing to pollute are greater than the costs of reducing that pollution. In this way, the tradeoff between economic growth and pollution could be incorporated into decisions made by individual businesses. If the advantages of the additional production were greater than the costs of the pollution, then that production would take place. And society would be compensated by the fee for the harm caused by the pollution.

Questions To Think About

• To what extent are regulations which allow a certain amount of pollution and no more based on evidence that is meaningful scientifically?

• At what point in the process should the decision be made that there is sufficient scientific evidence connecting an energy-related activity and an environmental problem to require government or industry action? Can such a decision be made with any certainty at all?

• What evidence is there that current environmental regulations have been successful in reducing environmental damage? The indicators include statistics on automobile emissions, carbon monoxide concentrations, and so on. How does progress in these areas compare with the costs of the regulations?

• Do proponents of emissions fees specify practical ways of determining how the size of those fees will be determined?

• At what point should the possibility that the burning of fossil fuels will alter global climates fifty to seventy years from now (see Chapter 6) influence current policy? More generally, how far into the future should we look in shaping current behavior and policy?

• How sensitive should environmental regulations be to the impact they have on specific industries, such as public utilities, steel companies, or automobile manufacturers? How sensitive should they be to their impact on the economy as a whole?

RESOLVING THE ROLE OF NUCLEAR POWER

Where an individual stands on the question of nuclear power safety can usually be determined by the answer given to three questions: How great are the odds on a meltdown of a reactor core and how serious the consequences? Is there a solution to the problem of nuclear waste disposal? And can the peaceful uses of nuclear power be adequately separated from its military uses?

On the question of a massive accident related to a meltdown of a reactor core, the antinuclear movement asserts that the danger has a sufficiently high probability to be taken seriously. Incidents such as the accident at Three Mile Island in 1979 show, they say, that the series of errors and malfunctions necessary for a major accident is, indeed, possible. And they note that some studies have estimated that thousands of deaths could be caused by such an accident.

Others argue that the dangers have been exaggerated and cite a series of studies concluding that the chances of such an accident occurring are extremely slim. In one case the odds of a meltdown of a reactor core were placed at 1 in 200 million per year of reactor operation. They argue, therefore, that the safety risks posed by nuclear power are far more remote than those which surround certain other energy sources, such as coal.

There are also widely varying assessments on the question of nuclear waste disposal. Those opposed to nuclear power voice the opinion that no proved means of safe disposal exist and that many of the technical and scientific problems of waste disposal are just beginning to be addressed. The exhaustion of temporary storage space, they say, should not dictate a premature rush to a permanent storage site. Thus there has been serious opposition to efforts to establish a national waste disposal site. This opposition to a permanent site irks advocates of nuclear power, who see their opponents as simply obstructionist and blame them for inventing a problem.

In contrast to the other two questions few suggest there is a technical "solution" to the problems of nuclear proliferation. Instead, some argue that a pervasive system of inspections by international agencies and restrictions on exports of sensitive materials can reduce the threat of nuclear proliferation to acceptable levels. Besides, they assert, a country can build a nuclear weapon easily enough without involving a nuclear power plant (see Chapter 7).

From the antinuclear perspective, inspections and export restrictions are inadequate tools to prevent countries from manufacturing nuclear bombs with material and knowledge gained from nuclear power plants. Inspections are infrequent and it is often impossible to tell if some of the plutonium has been taken out of the reactor for possible use elsewhere. And export restrictions are doomed to failure since companies and, on occasion, countries will find ways to get around them.

Questions To Think About

- Are the risks of nuclear power offset by other advantages, such as lower costs or its contribution as an alternative to fossil fuels?
- How do the environmental risks of nuclear power compare with those of coal, oil, and other energy sources used to generate electricity?
- What evidence is there that technological advances may provide an answer to questions about nuclear power safety? What progress is being made in providing a permanent waste disposal site and are attempts successful to develop nuclear fuels or processes that make manufacture of weapons-grade material more difficult? (Chapter 4 provides some discussion of these issues.)
- Does the small possibility of a catastrophic accident involving nuclear power merit greater attention and expenditure of resources than the higher probability of accidents involving other energy sources, such as coal or hydroelectricity?
- How should one interpret the halt in new nuclear plant orders and the cancellation of many orders already placed? Are they the result of costly safety regulations, high interest rates, escalating construction costs, reduced demand for electricity in general, or fears of unmanageable accidents? What do one or another of these factors imply for the future of nuclear power?
- How do those who advocate immediate closing down of nuclear power plants propose that the electricity needs of areas presently served by those plants be met and who would bear the resulting costs?

MANAGING RESOURCE SCARCITY

This issue is commonly approached through the question, When will we run out of...? To some, the answer is that the date of resource exhaustion is uncomfortably close if present patterns of consumption and production continue. They worry that an imperfectly functioning market will lead to overconsumption of resources such as oil and gas and that alternative energy technologies will not be ready to take the place of the conventional fuels be-cause of sluggish government action and misplaced corporate priorities. Thus, society will be forced to do without, and a sharp drop in the standard of living will result.

One response to this view is to question the trends outlined or the data they are based on. Frequently, assumptions about the availability of resources or consumption trends can be questioned. A different critique of the limits-to-growth position is to deny that the question, When will we run out of...? is, in fact, a useful one to ask, or that it is susceptible of being meaningfully answered. Holders of this latter view do not question that there is a finite store of physical resources which, at *some* date, could conceivably be exhausted. But the history of resource development shows that continued depletion of any exhaustible resource translates into higher cost, curbing demand and heightening incentives to develop less accessible, more costly, deposits, or al-ternatives. Typically, the increasing scarcity of a resource brings about such adjustments long before the commodity vanishes.

A counterargument is that, because past resource shifts occurred rela-tively smoothly, this is no assurance that history will repeat itself. For one thing, in contrast to the past, the environmental implications of alternative courses of action need to be considered. Second, the level of demand on these resources is now vastly greater. Even so, the problem is a far different one from picking out the day the world will run out of a resource.

Questions To Think About

- Is anxiety about depletion based on the assumption that demand will grow at a rate unchecked into the future?
- What is assumed about the price? Very few things are consumed along prior trends when they become more costly.
- What evidence is there that additional quantities of a scarce resource will be found? Indicators include trends in commitments of private capital as well as statistics on drilling, exploration, and development. Beyond this, new geologic knowledge and theories can help determine where de-posits are to be found. (Chapter 3 discusses the overall resource picture.)
- Is development of alternative energy sources impeded by public poli-cies—for example, distorted budgetary priorities, or the lobbying of spe-cial interest groups?
- Which alternative energy resources portend such a stretched-out devel-opment path that traditional private incentives—guided by reasonable expectation of commercial success—are insufficient? Which develop-ments can be expected to unfold within the traditional bounds of private-sector enterprise?

SUBSIDIZING ENERGY PRODUCTION

The necessity for subsidies in the energy area has never been apparent to those with the strongest faith in the free market. They see nothing positive resulting from this sort of government intervention. Any money spent on energy will have to be financed by increasing government deficits or reducing other spending programs. Subsidies distort the way in which the market chooses winners and losers and needlessly involve government in decisions about the profitability of various energy industries.

Others have less faith in the market and see broad needs which the government must meet through tax and spending policies. With a limited amount of government funds available for energy, a decision to provide a subsidy to consumption or production of a particular energy source implies that less money will be available for other resources. Thus, backers of a particular subsidy need to document an extra dividend which this subsidy will return. Often the rationale is national security. A backer of a particular energy subsidy will argue that government action is required in order to reduce oil imports. Others will argue that environmental concerns justify government action to subsidize a nonpolluting substitute. Still others will offer employment as a justification, noting that subsidies are necessary because society cannot allow the layoffs of workers tied to established industries.

A somewhat different argument is offered by those who agree that the market is not perfect but think its weaknesses are limited to certain, more narrowly defined areas. One such area is the "high-risk, long-term, high-payoff" ventures such as nuclear fusion or breeder reactor technology (see Chapter 4). It is suggested that a private investor cannot be expected to bear the large costs of projects which will only be economical after many years, if ever. Since these projects could be critically important, there is a rationale for government involvement.

Questions To Think About

- Does the advocate of a subsidy to one energy source explain persuasively why it is more worthy of special treatment than another source?
- Should government limit its role to making sure that all energy alternatives compete on even ground or should it afford special treatment to certain sources felt to be technologically or economically superior?
- Could a particular research and development activity that government is being asked to finance be funded with equal or greater success by private enterprise?
- Could a subsidy to a particular energy source actually slow down its commercial development by reducing the need for meeting competition through innovation?

- Considering budget outlays, "off budget" transactions, "tax subsidies," loan guarantees, and other forms of subsidization, which energy source is already receiving the most government help?
- How does subsidizing energy production compare with subsidizing energy conservation?

THE "HARD" VERSUS "SOFT" ENERGY PATH

For some people, the most critical aspect of future energy use in the United States is whether it pursues the "hard path" or the "soft path." The hard path is characterized by reliance on nonrenewable fossil fuels and nuclear reactors or other large-scale production facilities. In the future, the hard path is assumed to draw increasingly on electricity generated predominantly from coal or nuclear reactors. In contrast, the soft path emphasizes conservation, the use of decentralized, renewable energy sources which are considered "appropriate" to the task for which energy is required, and such energy-saving technologies as wood-burning boilers or furnaces, cogeneration of electricity (discussed in Chapter 2), and passive solar energy.

Advocates of the soft path favor this approach for several reasons. They say it will minimize pollution and avoid the threats to safety posed by nuclear power. Because it relies on fossil fuels only as a transition to a renewable energy future, society will not be threatened by resource depletion. And soft technologies are available without the capital costs or siting problems of nuclear or coal plants. Finally, supporters of the soft energy path say that most of the technologies involved will soon be or already are feasible economically and technically. The problem, they say, is that the energy market, dominated by large corporations, is extremely biased in favor of conventional technologies. This bias is reinforced by present government policies such as misplaced research and development priorities and the tax code. They therefore call for active government involvement to promote soft technologies and counterbalance corporate influence.

Those who support the hard energy alternative argue the soft energy path is costly and that allocating resources to it only diverts them away from more productive uses. They take issue with the idea that the world is running short of oil and gas, pointing to the recent increases in production in places such as the North Sea and Alaska. They are skeptical of claims that renewable energy alternatives will be economically feasible in the near future and accuse soft energy supporters of trying to impose their life-style on society.

A different perspective suggests that much of the hard–soft energy debate misses the point. The ideal energy source for a particular end use is the one which is least costly for the task, assuming that cost calculations take account of environmental and other relevant factors. If consumers attach sufficient value to having their homes heated by solar power, they will choose this alternative even if it has a higher price. The role of government is not to tilt the outcome unfairly toward one side or the other by means of its considerable financial clout. They admit that this may involve eliminating government-created cost advantages for many hard energy technologies.

Questions To Think About

- Do the prices for hard energy alternatives generally reflect their cost to society? What about prices of soft energy options? Are present price differences accounted for by implicit or explicit subsidies?
- Are consumers less likely to purchase renewable energy technology that has significant up-front costs but is economical over a longer time period? Does this suggest a need for government action?
- Are there institutional or political factors which persistently favor the interests of one energy resource industry over another? Are these factors biased toward "big ticket," hard energy items and suppliers?
- What social impacts would widespread implementation of the soft energy alternative have? Is it worth paying a premium to achieve these impacts?
- Should government efforts to spur solar energy development concentrate on research and development or on furthering the commercial success of technology already in existence?

The issues examined in this Epilogue may be but a small sampling of the energy questions that could surface in the next decade. But there is little doubt that these issues will be with us for the foreseeable future, that underlying conditions will change during that time, and that policy preferences will also change. No one can write a "once and for all" energy book. One only can hope to provide a guide for keeping up with the energy problem by identifying issues, encouraging the search for alternative ways of dealing with them, and assessing the probable consequences of different policy choices.

Glossary

Acid rain. Any precipitation, rain or snow, which is abnormally acidic. Such precipitation usually contains sulfuric or nitric acid and is blamed for pollution of lakes and damage to crops and trees. Coal burning, according to some, is the primary cause.

Active solar energy. Solar energy systems which use mechanical components, such as pumps, in order to produce energy.

Anthracite coal. A type of coal with high carbon and high heat content.

Base-load generating costs. The cost of electricity generated from a facility operating at high constant output with little hourly or daily fluctuation (see also **Peak-load generating costs**).

Biomass. A general term for living matter. In an energy context, that which contains the potential for heat or energy generation; plants, crops, sewage, and waste material are all sources of biomass energy.

Bituminous coal. Coal with less carbon than anthracite and a heat content which, compared with other types of coal, ranges from intermediate to high.

Bottled gas. Popular name for liquefied petroleum gas.

Authors' note: In the preparation of this glossary, substantial use was made of the glossary prepared at RFF for Energy: The Next Twenty Years *(Cambridge, Mass., Ballinger, 1979).*

Breeder reactor. A type of nuclear reactor that produces more fissile material than it burns.

British thermal unit (Btu). The amount of energy necessary to raise the temperature of one pound of water by one degree Fahrenheit, from 39.2 to 40.2 degrees Fahrenheit.

Bubble concept. Setting one standard for the total emissions of several polluting sources in a specific area, rather than separate standards for each one. This is designed to allow flexibility for meeting pollution control targets by making it possible to offset increased emissions from one source with reduced emissions from another nearby source.

Carbon dioxide problem. See **Greenhouse effect.**

Clinch River breeder reactor. A proposed demonstration project for a type of breeder reactor. The CRBR, located in Tennnessee, would operate on a plutonium fuel cycle. It has been the subject of considerable debate.

Coal conversion. Commonly used to denote coal gasification (conversion to gas) and liquefaction (conversion to liquid fuels).

Coal slurry pipeline. A pipeline which transports crushed coal mixed with a liquid, usually water.

Cogeneration. The production of both electricity and useful heat from a burning fuel, thereby increasing the efficiency with which fuel is used. Can be used in either factories or utility generating plants.

Common property resources. Resources owned by the entire society, such as oceans, rivers, and the air.

Concentration ratio. Measure of the degree to which an industry is dominated by a small group of firms. Usually expressed as the percentage of sales accounted for by the largest four or eight firms in a particular industry.

Conglomerate. A combination of firms under one management operating in more than one business. Can be as closely related as an oil company with subsidiaries in the chemical business, or as dissimilar as an oil company operating a retail chain.

Constant dollars. Dollar estimates from which the effects of changes in the general price level have been removed. Useful in comparing dollar values in different years, and reported in terms of a base-year value, such as 1972 dollars.

Conversion efficiency. The percentage of total thermal energy of a given resource that is actually converted into usable energy; frequently applied to percentage of thermal energy converted into electricity by an electric generating plant.

Crude oil. The unrefined state in which petroleum is found and brought to the surface.

Current dollars. Dollar values as they are commonly expressed, which have not been corrected for changes in the general price level (see also **Constant dollars**).

Declining block electricity rates. Electricity rates which become progressively lower on a per-kilowatt-hour basis as more electricity is consumed.

Decontrol. Removal of government price controls so that market forces determine the price of a product.

Depletion allowance. A tax allowance extended to the owner of an exhaustible resource presumed to reflect the reduction in value of the resource in place caused by extraction.

Deregulation. See **Decontrol**.

Devonian shale. Gas-bearing black or brown shale of the Devonian geologic age; underlies large area of the Appalachian Basin.

Discount rate. A percentage rate used to adjust future costs and benefits to their present day value.

Elasticity of demand. The percentage change in the quantity demanded that results from a percentage change in another economic variable, such as price or income. Demand is said to be price elastic if the percentage change in the amount of an item purchased is greater than a percentage change in price. It is price inelastic if the amount purchased does not react strongly to price movements.

Emissions charge. A monetary charge placed on polluters calculated per unit of pollutant discharged.

Enhanced oil recovery. Technology for lifting oil from wells beyond amount recovered through natural reservoir pressure. Commonly defined to exclude simple procedures such as injecting water into a well.

Enrichment. The process by which the percentage of the fissionable isotope uranium 235 is increased above that found in natural uranium. Necessary for use of uranium in most types of reactors.

Environmental Impact Statement (EIS). An analysis of the environmental effects of a proposed action. Under the U.S. National Environmental Policy Act of 1969, an EIS must be prepared in conjunction with any proposed action of the federal government that would significantly affect the quality of the environment.

Ethanol.   Also known as ethyl alcohol or grain alcohol. A flammable liquid which can be blended with or substituted for gasoline. Can be made from fermented or distilled liquors, sugarcane, or petroleum.

External costs.   Those costs which are imposed on a third party not engaged in an activity or involved in a transaction. An example would be air pollution caused by burning of fossil fuel to generate electricity. Neither the producer nor the buyer of electricity pay the costs created by the pollution.

Fissile material.   Atoms that fission upon the absorption of a low-energy neutron.

Fission.   The splitting of an atomic nucleus with the consequent release of energy. All present nuclear reactors utilize this process.

Flue gas desulfurization.   The removal of sulfur oxides from the gases emitted during fuel combustion before they reach the atmosphere. Most often used in conjunction with coal burning. Scrubbers can be used for this purpose.

Fluidized-bed combustion.   An unconventional process of burning coal in a bed of limestone-based particles which captures much of the sulfur and thus reduces its release into the atmosphere.

Fusion.   The combining of certain atomic nuclei, with a consequent release of energy. Potentially a virtually inexhaustible energy resource in reactors of the future.

Gasohol.   Either alcohol derived from an organic material (such as grain, sugar, or waste) or a mixture of such alcohol with gasoline. Both can be used as motor fuel.

Geopressured brines.   Water contained in some sedimentary rocks under abnormally high pressure; unusually hot and may be saturated with methane.

Geothermal energy.   The heat energy in the earth's crust which derives from the earth's molten interior. Can be tapped as steam, or by injection of water to form steam.

Gigawatt.   The unit by which the electric-generating capacity of a country is sometimes measured. One gigawatt represents 1 billion watts, 1 million kilowatts, or 1,000 megawatts. It is thus equal to the approximate size of a large power plant.

GNP deflator.   A price index that translates GNP in current dollars into GNP in constant dollars, thereby correcting for the change due to general price movements.

Greenhouse effect. A possible increase in the average temperature of the earth's atmosphere caused by an increase in the atmospheric concentration of carbon dioxide or other gases.

Half-life. The period required for the decay of half of the atoms in a given amount of a specific radioactive substance.

Heat pump. A device for transferring heat from a substance at one temperature to a substance at a higher temperature, by alternately vaporizing and liquefying a fluid through the use of a compressor. Electric heat pumps in residences extract heat from the outside air for winter heating and from the inside air for summer cooling.

Heavy oil. Crude oil of such low viscosity that it does not flow freely enough to be lifted from a reservoir reached by a drill.

Highly enriched uranium. Uranium with enough of the fissionable element isotope uranium 235 to be usable as an explosive. Natural uranium must be subject to certain mechanical or chemical processes before reaching this state.

Horizontal integration. Combination of two or more firms that are producing competitive products. For example, a coal producer that purchases another coal mine.

Indirect energy costs. The cost of the energy involved in the production, marketing, and distribution of goods. Consumers pay direct energy costs for items such as fuel oil or gasoline. They pay indirect energy costs for the energy embodied in the goods they purchase.

Kilowatt. 1,000 watts (see below). A unit of electrical power.

Kilowatt-hour. A measure of electricity use. A kilowatt-hour measures the amount of energy expended by one kilowatt in one hour.

Kinetic energy. The energy a body has because of its motion.

Light distillates. Distillate fuel derived from heating crude petroleum to a certain level. Includes naphtha and aviation fuel.

Light water reactor. The type of reactor most commonly in use today. It uses ordinary water as a coolant to transfer heat from the fissioning uranium to a steam turbine and employs slightly enriched uranium.

Lignite. The poorest quality of coal, it has low heat content, and a high percentage of volatile matter and moisture; an early stage of coal formation.

Liquefied natural gas. Natural gas that has been liquefied for purposes of storage or transportation.

Liquefied petroleum gas. Commonly known as bottled gas, it can be propane or butane or a mixture of the two which has been liquefied to facilitate storage and transportation.

Low-head hydroelectricity. Small-scale hydroelectric power in which the height from which the water falls to drive a turbine is less than large dams.

Marginal cost pricing. Charging users for all units consumed at the rate that corresponds to the cost of the final unit that needs to be supplied to meet demand. Sometimes referred to as "incremental" pricing.

Market structure. The organization of a particular industry or a segment of an industry. Refers to number of firms, their size, ease of entry, and so on.

Megawatt. The unit by which the electric-generating capacity of power plants is usually measured. The capacity of a 1,000-megawatt (MW) plant represents 1 million watts or 1,000 kilowatts (kW). Large plants typically measure between 500 and 1,000 MW.

Meltdown. A nuclear reactor accident in which the radioactive fuel overheats, melting the metal-encased fuel rods. Could result in release of radioactive gases and clouds of particulates into the environment.

Methane. The principal constituent of natural gas. Occurs naturally in petroleum reservoirs. The gas can also be derived from treatment of sewage, and animal and plant crop wastes. It has a high energy content and burns with a clean flame.

Middle distillates. A term used to describe the petroleum products resulting from boiling petroleum at temperatures between 200°C and 350°C. These include light fuel oil and diesel fuel.

Natural gas liquids. Hydrocarbon components of natural gas that are liquefied when natural gas reaches the wellhead. Propane, butane, and pentane are all natural gas liquids (these can also be manufactured at refineries).

Natural uranium. Uranium whose properties have not been artificially altered through processes such as enrichment.

Ocean thermal energy conversion. Power generation by exploiting the temperature difference between surface waters and ocean depths. Referred to as OTEC.

Oil shale. Rock containing organic matter (kerogen) that upon being heated to 800°F–1,000°F can yield commercially useful oil and gas.

OPEC. The organization of Petroleum Exporting Countries established in 1960. Its members in 1983 were Algeria, Ecuador, Gabon, Indonesia, Iran, Iraq, Kuwait, Libya, Nigeria, Qatar, Saudi Arabia, United Arab Emirates, and Venezuela.

Overthrust Belt. A geologic zone stretching from Alaska to Mexico that is thought to be the site of substantial oil and gas reserves. Recent finds have been centered in Wyoming, Utah, and Idaho.

Ozone. A gas found in both the upper and lower atmosphere which can be created by the interaction between volatile organic compounds and nitrogen oxides in the presence of sunlight. It is the principal component of smog. Ozone in the upper atmosphere helps shield living organisms from damaging ultraviolet radiation.

Passive solar energy. Means of utilizing solar energy for heating and cooling, such as location, window shutters, eaves, or insulation, that do not consume electricity, pump fluids, or have powered mechanical components.

Peak-load generating costs. The cost of electricity from a facility operating at its highest possible output, usually in extreme weather conditions. Peak-load costs often reflect the costs of operating expensive auxiliary generating plants.

Petrochemicals. Any chemical which is derived from petroleum products.

pH. A measure of acidity, calibrated logarithmically, whose values run from zero to 14, with 7 representing neutrality, numbers less than 7 increasing acidity, and numbers greater than 7 increasing alkalinity.

Photovoltaic cells. Devices that convert sunlight directly into electricity.

Plutonium. A long-lived radioactive metallic element not found in nature but formed in nuclear reactors. Can be used to yield energy within nuclear reactors, primarily breeder reactors, and is the typical constituent of nuclear bombs.

Primary energy. Energy before processing or conversion into different or more refined form.

Process heat. Heat used for an industrial operation or process.

Propane. A gaseous hydrocarbon occurring naturally in crude petroleum and natural gas liquids. It is readily liquefied and can be used as a convenient heating and cooking fuel.

Quad. A quadrillion Btus (see conversion table on page 226).

Reactor core. The region within a nuclear reactor that contains the fissionable material.

Real dollars. See **Constant dollars.**

Renewable energy sources. A term applied to energy sources which are not subject to depletion and exhaustion. These include solar energy, wind, wave, hydroelectric power, and, in a broad sense, biomass.

Reprocessing. The chemical and mechanical processes by which useful plutonium and uranium are recovered from spent reactor fuel.

Reserves. Resources well identified as to location and size and commercially producible at current prices and with current technology.

Residual fuel oil. The constituents of petroleum remaining after lighter products such as gasoline, kerosine, and diesel oil have been distilled. Typically used as industrial and utility boiler fuel.

Resources. Those portions of a given energy source estimated to remain discoverable and recoverable in the future; sometimes used to denote the sum total of a given resource estimated to exist.

Rolled-in-pricing. Injecting a high-cost product into the market by averaging its cost with that of the bulk of the same product produced at lower cost. Typical for public utility marketing of gas and electricity.

Royalties. Payments made by an oil producer to owners of land from which oil is extracted. These are paid upon production and usually calculated on a per-barrel basis.

Rural electric cooperatives. Member-owned organizations which deliver electricity to rural areas. Many receive government and private financing.

Saudi marker crude. A type of Saudi Arabian light crude oil used as the standard for measuring prices of other grades of oil.

Scrubbing. Techniques for removing pollutants from gas emissions resulting from burning of fossil fuels.

Shale oil. See **Oil shale.**

Spent fuel. The fuel elements removed from a reactor after several years of generating power. Spent fuel contains radioactive waste materials, and unburned uranium and plutonium. May be reprocessed, stored temporarily, or disposed of permanently.

Spot price. Price formed in a market for one time, usually quick delivery of a specific cargo to a specific destination. In times of stringency, spot market prices can greatly exceed prices for identical products sold under long-term contracts. When demand is depressed, spot market prices are typically below contract prices.

Strategic Petroleum Reserve (SPR). The government-owned U.S. crude oil stockpile, presently centered in Louisiana salt domes. Plans are to use this reserve in an emergency.

Stripper wells. Oil wells producing less than 10 barrels per day.

Synthetic fuels. A term now commonly reserved for liquid and gaseous fuels that are the product of a conversion process rather than mining or drilling; most frequently applied to cover liquids or gases derived from coal, shale, tar sands, waste, or biomass.

Tar sands. Formations of loose-grained rock material bonded by a heavy tarlike substance. Can be surface-mined and processed to yield a liquid with the properties of crude oil. In 1982 there was commercial production of tar sands in Canada.

Thorium. A radioactive metallic element. Can be made fissile in breeder reactors and is a possible alternative to uranium as a reactor fuel.

Threshold. A dose level or concentration (as of radiation or pollutant) below which there is, or it is assumed there is, no effect on a recipient tissue or organism.

Tidal energy. Energy derived from exploiting the rise and fall in the level of coastal waters.

Tight sands. Gas-bearing geologic areas holding gas so tightly that bringing it to surface by conventional extraction methods is impossible without special stimulation.

Unconventional gas. Gas developed from uncommon sources such as tight sands, black shale, geopressured zones, and coal seams. Usually requires costly special reservoir engineering techniques to permit recovery. Although there may be large potential amounts of energy in these sources, they have not been widely exploited.

Uranium. A radioactive metallic element that is used to fuel nuclear reactors.

Uranium oxide. The most common oxide of uranium found in typical ores. Also, the unit of measurement in which uranium resources and requirements are commonly calculated and reported.

Vertical integration. Expansion of a single firm into operation at more than one stage of an industry, such as an oil company engaging in production, refining, and marketing.

Watt. A unit of power equivalent to one ampere under an electrical pressure of one volt. Signifies the capacity to produce electricity.

Weapons grade material. Any mixture of uranium or plutonium isotopes that could be used to create a nuclear explosion.

Wellhead price. The price of oil (or gas) as it reaches the surface at the top of the production well.

Wellhead tax. A tax imposed on oil (or gas) at the top of the production well.

Windfall profits tax. A tax on profits that do not arise from what are considered normal operations. The tax imposed by the U.S. government in 1979 on crude oil producers to reduce profits from price decontrol was labeled a windfall profits tax.

Conversion Table

Our understanding of energy issues is sometimes impeded by a multitude of measurement units—ranging from the familiar million barrels per day to less common terms such as quads, megajoules, and the dreaded kilocalories. It is not necessary for the average reader to keep in mind the formulas for converting one unit of measure to another. But it helps to be familiar with some of the units commonly found in energy literature.

Because a variety of resources can be used to produce energy, a measure which can be used for all of them is necessary. This is the role of the British thermal unit (Btu), defined as the amount of energy required to raise the temperature of 1 pound of water by 1° Fahrenheit, that is, from 39.2° F to 40.2° F. The accompanying table has been divided into three parts. The upper panel is designed to show first, what a Btu is in terms of the common measurements of use of the primary energy sources, and, second, what a quadrillion (1,000,000,000,000,000) Btu—in energy jargon, a "quad"—is equivalent to. The conversion table in the center shows different ways of portraying the equivalent of 1 million barrels per day of oil. The lower portion of the table gives electricity conversions. All values are approximate.

1 barrel of crude oil (42 gallons) = 5,800,000 Btu
1 short ton of bituminous coal (2,000 pounds) = 23,150,000 Btu
1 cubic foot of natural gas = 1,000 Btu

1 quadrillion Btu equals:
170 million barrels of crude oil (470,000 barrels per day)
43 million short tons of coal
975 billion cubic feet of natural gas

1 million barrels per day of crude oil equals:
2.1 quadrillion Btu
91 million short tons of coal
2 trillion cubic feet of natural gas

Electricity
1,000 watts = 1 kilowatt
1 kilowatt-hour (kWh) = the amount of energy expended by 1 kW in one hour; the heat value
of 1 kWh is 3,412 Btu. Producing a kilowatt-hour requires about 10,000 Btu of fuel.

Index